Differentiated M

Sprints 5

Blackline Masters

Tricia Salerno of SMARTTraining LLC

Singaporemath.com Inc®

SingaporeMath.com Inc®

Copyright © 2010 SingaporeMath.com Inc.

Published by
SingaporeMath.com Inc
404 Beavercreek Road #225
Oregon City, OR 97045
U.S.A.
E-mail: customerservice@singaporemath.com
www.singaporemath.com

First published 2010 in the United States of America

Written by
Tricia Salerno
SMARTTraining LLC
www.SingaporeMathTraining.com

Cover design by
Jopel Multimedia

Differentiated Math Sprints (Blackline Masters) Level 5
ISBN: 978-1-932906-45-5

Printed in the U.S.A.

Introduction

"They don't know their facts!" This is the common lament we hear from teachers around the United States. We've all tried flash cards, card games, asked for help from home, and many of our students still don't know their facts.

This book is here to help. Contained herein are math activities called "Sprints." When administered correctly, students **love** to take sprints. Yes, your students will ask you, "Do we get to take a sprint today?" Imagine students ASKING to practice their math facts. Sound impossible? Read on.

A sprint is a timed math test for FUN! That's right…**no grades** are ever taken on sprints. In fact, the students should not put their names on their papers. The clear statement to students that these are not for a grade should alleviate any math anxiety which sometimes arises during timed tests.

The teacher administering the sprint acts as if the students are actually involved in a race. The excitement is palpable in the classroom as the students get their pencils ready. "On your mark, get set, GO!" Students race to beat their own scores as they complete as many problems as they can in 60 seconds.

If the teacher models excitement, and keeps up the level of excitement through the year, the results are astounding.

Importance of Math Facts

The importance of automatic recall of basic math facts has been argued in the past. In this day of technology, some say, why is it important to know the product of 6 and 8 when you can press a few buttons and have the answer quickly? In fact, some of the parents of our current students grew up with calculators in their hands. Those parents have no idea how to help their children with mastering math facts because they don't know the facts themselves.

One of the problems with lack of automaticity with math facts is that if too much mental energy has to be spent recalling a basic fact, there's no mental energy left to solve the problem. Gersten and Chard stated:

> "Researchers explored the devastating effects of the lack of automaticity
> in several ways. Essentially they argued that the human mind has a

limited capacity to process information, and if too much energy goes into figuring out what 9 plus 8 equals, little is left over to understand the concepts underlying multi-digit subtraction, long division, or complex multiplication." Gersten, R. and Chard, D. Number sense: Rethinking arithmetic instruction for students with mathematical disabilities. *Journal of Special Education* (1999), 3, 18–29 (1999).

Importance of Mental Math

Mental math is important for many reasons. Cathy Seeley, former president of the NCTM, stated:

> "Mental math is often associated with the ability to do computations quickly, but in its broadest sense, mental math also involves conceptual understanding and problem solving….Problem solving continues to be a high priority in school mathematics. Some argue that it is the most important mathematical goal for our students. Mental math provides both tools for solving problems and filters for evaluating answers. When a student has strong mental math skills, he or she can quickly test different approaches to a problem and determine whether the resulting path will lead toward a viable solution." (*NCTM News Bulletin*, December 2005).

Adrenaline

Research has proven that adrenaline aids memory. James McGaugh, a Professor of neurobiology at the University of California at Irvine, proved that adrenaline makes our brains remember better. When a sprint is given with a sense of urgency, students experience a rush of adrenaline, and this aids their memory of the mental math being tested. It also makes the exercise significantly more fun!

About this Book

The Singapore Math curriculum sold by Singaporemath.com stresses the use of mental math. These sprint books are particularly useful to teachers using that curriculum. In fact, sprints are useful to all elementary teachers interested in developing mental math fluency in their students.

The section on the next page explains how to give a sprint. You will see that students are racing to beat their own scores each time they take one of these tests. In the excitement of the sprint, applause is given to the student who gets the most problems correct on the first half, and to the student who improves the most on the second half. This applause, is, of course, well-meaning. But what about those students who rarely, if ever, receive applause? How can we keep them motivated to take the sprints when the same few children are always receiving the applause? Equally important is another concern: how do we keep the strongest math students challenged taking sprints with their classmates?

This book is written to address these concerns. Each sprint is differentiated. The A sheet of each half of the sprint is for the majority of the class. The B sheet is for the strongest students. If you look closely at the A and B sheets of each sprint, the answers to the problems are the same. Many of the problems on the B sheets, however, require more mental calculation. These sprints will allow access to the excitement of sprints for all of your students.

If you teach in a fairly homogeneous classroom, using differentiated sprints is not important. In that case, use the A versions of both halves for one sprint for all of your students, and perhaps, later in the year, use the B versions. Bonus…this will give you double the number of sprints!

Generally speaking, sprints are used in the classroom two to three times per week. Once a teacher becomes experienced administering a sprint, the entire process, start to finish, should take about 10 minutes. You can re-use sprints. For example, if you give sprint 305 one week, you can give it again a couple of weeks later. The students will not remember it.

You may want to buy a sprint book at a grade below the level you teach so that students get used to taking sprints and feel very successful with them. Particularly if their mental math fluency is not where it should be, you can help them build it gradually by starting at a lower grade level.

How to Give a Sprint

1. Determine which sprint you want to give by looking at the topic of each. If you want to give sprint 305, in a differentiated classroom copy 305A and B first half and 305 A and B second half, remembering that B is only for your strongest students.

2. Hand out, face-down, a "First Half" sheet for students to attempt to complete in 1 minute. The strongest students should receive the B sheet of the first half. All other students receive the A sheet. Instruct students not to turn the sheet face-up until told to "GO!" Get students excited and enthuse: "On your mark, get set, GO!" and start your timer.

3. When the timer rings indicating one minute has elapsed, instruct students to:

    a) stop working
    b) draw a line under the last problem they have completed
    c) put pencils down.

4. Read the correct answers while students pump their hand in the air and respond "**yes**" to each problem that was answered correctly. They should be silent for any incorrect answers. Tell students to mark the number of correct problems at the top of the page. Ask how many students got at least one right on their "sprint." (All hands should raise.) Two right? Three? Continue until there is only one hand left in the air and applaud for that person.

5. Tell students to take a couple of minutes to calmly and quietly try to complete the worksheet.

6. After a couple of minutes, tell students to stand-up, push in their chairs and make sure they have some space.

7. Then lead them in chorally skip counting (10, 20, 30,etc.) while completing a simple physical exercise. Jumping jacks, windmills, etc. are all fine. There is a theory that crossing the mid-line of the body while exercising is beneficial. For this reason, an exercise known as cross crawl is a good one. In this exercise, students raise their left knee to their right elbow, then the right knee to the left elbow.

8. Hurry students back to their seats and hand them the "Second Half" of the same sprint, face-down, once again giving B sheets to the strongest students.

9. Tell students that their goal in this second half of the sprint is to beat their first score by at least one. **They are competing only with themselves.**

10. Repeat the preceding procedure through step 4, except that after making the correct number of problems at the top of the page, they are to compare their scores on the first half to their scores on the second half.

11. Now ask, "How many of you correctly answered at least one question more on the second half of the sprint?" Hands should all raise. "Two better? Keep your hands up…Three better?... four?" and so on until only one person has their hand in the air. Applaud for the person who was the "most improved."

12. Instruct participants to toss both halves of the sprint in the recycling. It is important to let students know that this is simply a competition against oneself to improve mental math skills and it is for fun. It is NOT for a grade. If a student wants to bring the sprint home to show his or her parents, that's all right, but it should be the student's choice.

For a video of a sprint being given in a second grade classroom, look under the "Resources" tab at http://singaporemathtraining.com.

A good sprint:

1. Consists of two halves which test the same ONE skill. These are NOT random facts.
2. Builds in difficulty.
3. Is challenging enough that no one will be able to finish the first half in a minute.

NOTE: Look at each sprint and determine if your particular students can finish each half of a sprint in less than a minute. Some of the sprints have fewer problems than others. There is nothing wrong with doing a 30-second or 45-second sprint. Feel free to adjust the timing for your students, but be sure to keep the sense of urgency.

Acknowledgments

This series of books is due to the assistance of many people. Special thanks go to Ben Adler, Sam Adler, Laina Salerno, T. J. Salerno and Charlotte West for their hours spent taking and re-taking the sprints contained herein. Linda West made it all come together. Thank you, thank you, thank you.

Math Sprints 5

Round to the nearest million.

1.	1,220,000		15.	63,547,911
2.	5,369,942		16.	54,911,001
3.	9,496,351		17.	79,694,112
4.	6,973,422		18.	98,621,496
5.	3,621,913		19.	100,521,647
6.	4,474,991		20.	99,546,381
7.	9,999,999		21.	206,057,292
8.	7,248,645		22.	362,943,681
9.	10,306,942		23.	551,364,989
10.	12,399,999		24.	499,862,139
11.	26,842,396		25.	398,962,462
12.	19,842,396		26.	589,642,111
13.	39,512,362		27.	999,099,999
14.	9,459,583		28.	846,109,534

Math Sprints 5

1.	1,120,000	15.	52,547,911
2.	4,369,942	16.	72,911,001
3.	6,496,351	17.	84,103,999
4.	7,973,422	18.	20,621,496
5.	3,621,913	19.	100,599,999
6.	5,476,999	20.	99,521,310
7.	9,999,999	21.	201,057,292
8.	8,396,241	22.	299,943,681
9.	13,306,942	23.	551,364,989
10.	14,999,999	24.	599,862,139
11.	35,742,396	25.	798,962,462
12.	21,842,396	26.	649,642,111
13.	39,512,362	27.	999,099,999
14.	17,459,583	28.	709,609,534

Round the sum to the nearest million.

1.	1,243,601 + 10,000	15.	54,324,309 + 10,000,600
2.	4,962,349 + 100,000	16.	44,400,110 + 11,000,000
3.	9,043,999 + 340,000	17.	37,552,092 + 42, 000,000
4.	6,137,481 + 800,000	18.	34,510,752 + 64,000,000
5.	3,310,912 + 310,000	19.	49,611,214 + 50,900,000
6.	4,376,211 + 100,011	20.	45,923,081 + 54,000,000
7.	9,462,964 + 600,000	21.	103,834,182 + 102,000,000
8.	5,124,324 + 2,100,000	22.	140,921,371 + 222,000,000
9.	5,263,751 + 5,100,000	23.	100,500,000 + 450,789,150
10.	10,164,999 + 2,200,000	24.	385,599,568 + 114,000,000
11.	12,511,263 + 14,000,000	25.	299,581,231 + 99,000,000
12.	12,411,263 + 8,000,000	26.	295,000,000 + 294,964,521
13.	28,307,231 + 12,000,000	27.	545,054,545 + 454,000,004
14.	3,289,345 + 6,000,238	28.	596,954,968 + 249,064,521

| 501 B | Round the sum to the nearest million. | Second Half |

1.	1,143,601 + 10,000	15.	51,324,309 + 1,200,000
2.	3,962,349 + 100,000	16.	31,400,000 + 41,560,000
3.	6,043,999 + 340,000	17.	42,101,999 + 42,000,000
4.	7,137,481 + 800,000	18.	11,310,752 + 9,500,000
5.	3,310,912 + 310,000	19.	49,711,000 + 50,900,000
6.	5,276,211 + 200,000	20.	45,801,081 + 54,000,000
7.	9,999,999 + 1	21.	199,834,182 + 1,000,000
8.	6,124,324 + 2,100,000	22.	188,921,371 + 111,000,000
9.	10,263,751 + 3,100,000	23.	100,500,000 + 450,789,150
10.	14,000,000 + 900,000	24.	485,599,568 + 114,000,000
11.	21,411,000 + 14,900,000	25.	599,581,231 + 199,000,000
12.	13,511,263 + 8,000,000	26.	414,000,000 + 235,964,521
13.	28,307,231 + 12,000,000	27.	545,054,545 + 454,000,004
14.	9,289,345 + 8,000,000	28.	502,354,968 + 207,304,521

1.	1,000,000	15.	64,000,000
2.	5,000,000	16.	55,000,000
3.	9,000,000	17.	80,000,000
4.	7,000,000	18.	99,000,000
5.	4,000,000	19.	101,000,000
6.	4,000,000	20.	100,000,000
7.	10,000,000	21.	206,000,000
8.	7,000,000	22.	363,000,000
9.	10,000,000	23.	551,000,000
10.	12,000,000	24.	500,000,000
11.	27,000,000	25.	399,000,000
12.	20,000,000	26.	590,000,000
13.	40,000,000	27.	999,000,000
14.	9,000,000	28.	846,000,000

501 A & B   Answer Sheet   Second Half

1.	1,000,000	15.	53,000,000
2.	4,000,000	16.	73,000,000
3.	6,000,000	17.	84,000,000
4.	8,000,000	18.	21,000,000
5.	4,000,000	19.	101,000,000
6.	5,000,000	20.	100,000,000
7.	10,000,000	21.	201,000,000
8.	8,000,000	22.	300,000,000
9.	13,000,000	23.	551,000,000
10.	15,000,000	24.	600,000,000
11.	36,000,000	25.	799,000,000
12.	22,000,000	26.	650,000,000
13.	40,000,000	27.	999,000,000
14.	17,000,000	28.	710,000,000

502 A		Multiply.		First Half

1.	$10 \times 2 =$	15.	$700 \times 30 =$
2.	$20 \times 10 =$	16.	$50 \times 30 =$
3.	$3 \times 10 =$	17.	$50 \times 60 =$
4.	$23 \times 10 =$	18.	$50 \times 600 =$
5.	$8 \times 100 =$	19.	$30 \times 120 =$
6.	$80 \times 10 =$	20.	$300 \times 12 =$
7.	$50 \times 2 =$	21.	$8,000 \times 70 =$
8.	$50 \times 20 =$	22.	$200 \times 450 =$
9.	$45 \times 2 =$	23.	$110 \times 50 =$
10.	$45 \times 20 =$	24.	$600 \times 70 =$
11.	$6 \times 20 =$	25.	$120 \times 500 =$
12.	$60 \times 20 =$	26.	$37 \times 200 =$
13.	$60 \times 100 =$	27.	$5,600 \times 30 =$
14.	$450 \times 2 =$	28.	$8,000 \times 50 =$

Math Sprints 5

1.	$10 \times 1 =$	15.	$400 \times 30 =$
2.	$10 \times 10 =$	16.	$50 \times 30 =$
3.	$5 \times 10 =$	17.	$50 \times 60 =$
4.	$15 \times 10 =$	18.	$70 \times 600 =$
5.	$4 \times 100 =$	19.	$70 \times 300 =$
6.	$40 \times 10 =$	20.	$700 \times 30 =$
7.	$50 \times 8 =$	21.	$3,000 \times 90 =$
8.	$50 \times 80 =$	22.	$200 \times 450 =$
9.	$45 \times 2 =$	23.	$110 \times 30 =$
10.	$45 \times 20 =$	24.	$300 \times 80 =$
11.	$7 \times 20 =$	25.	$120 \times 500 =$
12.	$70 \times 20 =$	26.	$12 \times 600 =$
13.	$90 \times 100 =$	27.	$5,600 \times 30 =$
14.	$450 \times 2 =$	28.	$9,000 \times 50 =$

	502 B		Divide.		First Half

1.	$200 \div 10 =$		15.	$4{,}200{,}000 \div 200 =$
2.	$2{,}000 \div 10 =$		16.	$300{,}000 \div 200 =$
3.	$300 \div 10$		17.	$600{,}000 \div 200 =$
4.	$2{,}300 \div 10 =$		18.	$600{,}000 \div 20 =$
5.	$8{,}000 \div 10 =$		19.	$72{,}000 \div 20 =$
6.	$80{,}000 \div 100 =$		20.	$720{,}000 \div 200 =$
7.	$2{,}000 \div 20 =$		21.	$11{,}200{,}000 \div 20 =$
8.	$20{,}000 \div 20 =$		22.	$8{,}100{,}000 \div 90 =$
9.	$180 \div 2 =$		23.	$5{,}500{,}000 \div 1{,}000 =$
10.	$18{,}000 \div 20 =$		24.	$8{,}400{,}000 \div 200 =$
11.	$240 \div 2 =$		25.	$3{,}600{,}000 \div 60 =$
12.	$12{,}000 \div 10 =$		26.	$148{,}000 \div 20 =$
13.	$18{,}000 \div 3 =$		27.	$3{,}360{,}000 \div 20 =$
14.	$27{,}000 \div 30 =$		28.	$32{,}000{,}000 \div 80 =$

502 B   Divide.   Second Half

1.	100 ÷ 10 =	15.	3,600,000 ÷ 300 =
2.	1,000 ÷ 10 =	16.	300,000 ÷ 200 =
3.	500 ÷ 10	17.	600,000 ÷ 200 =
4.	1,500 ÷ 10 =	18.	840,000 ÷ 20 =
5.	4,000 ÷ 10 =	19.	840,000 ÷ 40 =
6.	40,000 ÷ 100 =	20.	8,400,000 ÷ 400 =
7.	8,000 ÷ 20 =	21.	5,400,000 ÷ 20 =
8.	80,000 ÷ 20 =	22.	810,000 ÷ 90 =
9.	180 ÷ 2 =	23.	3,300,000 ÷ 1,000 =
10.	18,000 ÷ 20 =	24.	4,800,000 ÷ 200 =
11.	280 ÷ 2 =	25.	3,600,000 ÷ 60 =
12.	28,000 ÷ 10 =	26.	144,000 ÷ 20 =
13.	27,000 ÷ 3 =	27.	3,360,000 ÷ 20 =
14.	27,000 ÷ 30 =	28.	13,500,000 ÷ 30 =

1.	20	15.	21,000
2.	200	16.	1,500
3.	30	17.	3,000
4.	230	18.	30,000
5.	800	19.	3,600
6.	800	20.	3,600
7.	100	21.	560,000
8.	1,000	22.	90,000
9.	90	23.	5,500
10.	900	24.	42,000
11.	120	25.	60,000
12.	1,200	26.	7,400
13.	6,000	27.	168,000
14.	900	28.	400,000

Math Sprints 5

1.	10	15.	12,000
2.	100	16.	1,500
3.	50	17.	3,000
4.	150	18.	42,000
5.	400	19.	21,000
6.	400	20.	21,000
7.	400	21.	270,000
8.	4,000	22.	90,000
9.	90	23.	3,300
10.	900	24.	24,000
11.	140	25.	60,000
12.	1,400	26.	7,200
13.	9,000	27.	168,000
14.	900	28.	450,000

503 A		Solve.		**First Half**

1.	$3 \times 4 + 2 =$		11.	$(19 - 16) \times 8 =$
2.	$3 \times (4 + 2) =$		12.	$5 \times (8 + 9) =$
3.	$3 + 4 \times 2 =$		13.	$(6 + 8) \times 3 =$
4.	$(3 + 4) \times 2 =$		14.	$50 - (12 \times 4) =$
5.	$(6 - 4) \times 10 =$		15.	$53 - (12 \times 4) =$
6.	$(6 + 3) \times 7 =$		16.	$38 + (100 \div 5) =$
7.	$(15 - 3) \times 2 =$		17.	$(2 \times 3) + (2 \times 15) =$
8.	$2 \times (9 + 9) =$		18.	$2 \times (3 + 15) =$
9.	$24 \div (18 - 14) =$		19.	$(2 \times 64) - 49 =$
10.	$(5 + 3) + 5 =$		20.	$18 + (28 + 12) \div 4 =$

503 A   Solve.   Second Half

1.	$3 \times 4 + 1 =$	11.	$(19 - 16) \times 8 =$
2.	$3 \times (4 + 1) =$	12.	$7 \times (3 + 7) =$
3.	$1 + 4 \times 3 =$	13.	$(8 + 9) \times 5 =$
4.	$(1 + 3) \times 2 =$	14.	$60 - (7 \times 7) =$
5.	$(7 - 3) \times 5 =$	15.	$63 - (7 \times 7) =$
6.	$15 - 3 \times 5 =$	16.	$38 + (100 \div 5) =$
7.	$(15 - 3) \times 5 =$	17.	$(4 \times 2) + (35 \div 5) =$
8.	$2 \times (9 + 9) =$	18.	$3 \times (3 + 12) =$
9.	$32 \div (20 - 16) =$	19.	$(2 \times 32) - 19 =$
10.	$(5 + 6) + 5 =$	20.	$(99 - 27) \div 12 =$

503 B Solve. First Half

1.	$3 \times 4 + 2 =$	11.	$(21 \div 7) \times 8 =$
2.	$3 \times (4 + 2) =$	12.	$5 \times (4 + 4 + 9) =$
3.	$2 + 3 \times 3 =$	13.	$(6 + 4 + 4) \times 3 =$
4.	$(9 - 2) \times 2 =$	14.	$74 - (8 \times 9) =$
5.	$(9 - 5) \times 5 =$	15.	$77 - (8 \times 9) =$
6.	$(18 - 11) \times 9 =$	16.	$100 - (6 \times 7) =$
7.	$(50 - 2) \div 2 =$	17.	$(2 \times 5) + (2 \times 13) =$
8.	$2 \times (27 - 9) =$	18.	$2 \times (5 + 13) =$
9.	$(22 + 2) \div (18 - 14) =$	19.	$2 \times (50 + 14) - 49 =$
10.	$(50 - 11) \div 3 =$	20.	$18 + (6 + 17 \times 2) \div 4 =$

503 B   Solve.   Second Half

1.	$3 \times 4 + 1 =$	11.	$(21 \div 7) \times 8 =$
2.	$3 \times (4 + 1) =$	12.	$7 \times (4 + 4 + 2) =$
3.	$(2 \times 3) + 7 =$	13.	$5 \times (9 + 4 + 4) =$
4.	$(1 + 3) \times 2 =$	14.	$50 - (13 \times 3) =$
5.	$(9 - 5) \times 5 =$	15.	$53 - (13 \times 3) =$
6.	$18 - (9 \times 2) =$	16.	$100 - (6 \times 7) =$
7.	$(110 + 10) \div 2 =$	17.	$(32 \div 4) + (35 \div 5) =$
8.	$2 \times (27 - 9) =$	18.	$3 \times (3 + 12) =$
9.	$(31 + 1) \div (20 - 16) =$	19.	$2 \times (20 + 12) - 19 =$
10.	$(46 - 42) \times 4 =$	20.	$(51 + 9 + 3 \times 4) \div 12 =$

503 A & B		Answer Sheet		First Half

1.	14	11.	24
2.	18	12.	85
3.	11	13.	42
4.	14	14.	2
5.	20	15.	5
6.	63	16.	58
7.	24	17.	36
8.	36	18.	36
9.	6	19.	79
10.	13	20.	28

1.	13	11.	24
2.	15	12.	70
3.	13	13.	85
4.	8	14.	11
5.	20	15.	14
6.	0	16.	58
7.	60	17.	15
8.	36	18.	45
9.	8	19.	45
10.	16	20.	6

504 A		Multiply or divide.		First Half

1.	$25 \times 3 =$	11.	$52 \times 4 =$
2.	$25 \times 30 =$	12.	$52 \times 40 =$
3.	$70 \times 2 =$	13.	$52 \times 39 =$
4.	$4 \times 2 =$	14.	$50 \times 16 =$
5.	$74 \times 2 =$	15.	$45 \times 20 =$
6.	$74 \times 20 =$	16.	$80 \div 5 =$
7.	$28 \times 2 =$	17.	$117 \div 9 =$
8.	$28 \times 20 =$	18.	$750 \div 25 =$
9.	$28 \times 22 =$	19.	$600 \div 40 =$
10.	$21 \times 28 =$	20.	$2,040 \div 4 =$

504 A	Multiply or divide.	Second Half

1.	$30 \times 4 =$	11.	$32 \times 4 =$
2.	$5 \times 4 =$	12.	$32 \times 40 =$
3.	$35 \times 4 =$	13.	$32 \times 39 =$
4.	$6 \times 2 =$	14.	$25 \times 30 =$
5.	$80 \times 2 =$	15.	$25 \times 32 =$
6.	$86 \times 2 =$	16.	$78 \div 6 =$
7.	$15 \times 2 =$	17.	$750 \div 25 =$
8.	$15 \times 20 =$	18.	$117 \div 9 =$
9.	$15 \times 22 =$	19.	$900 \div 60 =$
10.	$21 \times 15 =$	20.	$4,921 \div 7 =$

504 B		Multiply or divide.	**First Half**

1.	$3 \times 25 =$	11.	$26 \times 8 =$
2.	$25 \times 30 =$	12.	$26 \times 80 =$
3.	$14 \times 5 \times 2 =$	13.	$26 \times 78 =$
4.	$2 \times 2 \times 2 =$	14.	$32 \times 25 =$
5.	$2 \times 74 =$	15.	$25 \times 36 =$
6.	$74 \times 2 \times 10 =$	16.	$96 \div 6 =$
7.	$14 \times 4 =$	17.	$143 \div 11 =$
8.	$14 \times 40 =$	18.	$750 \div 25 =$
9.	$14 \times 44 =$	19.	$600 \div 40 =$
10.	$42 \times 14 =$	20.	$1,530 \div 3 =$

504 B	Multiply or divide.		Second Half

1.	$4 \times 30 =$	11.	$16 \times 8 =$
2.	$4 \times 5 =$	12.	$16 \times 80 =$
3.	$35 \times 2 \times 2 =$	13.	$16 \times 78 =$
4.	$3 \times 2 \times 2 =$	14.	$30 \times 25 =$
5.	$2 \times 2 \times 40 =$	15.	$25 \times 32 =$
6.	$43 \times 2 \times 2 =$	16.	$91 \div 7 =$
7.	$3 \times 5 \times 2 =$	17.	$750 \div 25 =$
8.	$3 \times 5 \times 20 =$	18.	$143 \div 11 =$
9.	$15 \times 22 =$	19.	$900 \div 60 =$
10.	$21 \times 15 =$	20.	$8,436 \div 12 =$

504 A & B   Answer Sheet   First Half

1.	75	11.	208
2.	750	12.	2080
3.	140	13.	2028
4.	8	14.	800
5.	148	15.	900
6.	1480	16.	16
7.	56	17.	13
8.	560	18.	30
9.	616	19.	15
10.	588	20.	510

1.	120	11.	128
2.	20	12.	1280
3.	140	13.	1248
4.	12	14.	750
5.	160	15.	800
6.	172	16.	13
7.	30	17.	30
8.	300	18.	13
9.	330	19.	15
10.	315	20.	703

1.	$\dfrac{2}{4}$	16.	$\dfrac{10}{8}$
2.	$\dfrac{3}{9}$	17.	$\dfrac{26}{10}$
3.	$\dfrac{3}{18}$	18.	$\dfrac{42}{6}$
4.	$\dfrac{4}{20}$	19.	$\dfrac{15}{9}$
5.	$\dfrac{6}{9}$	20.	$\dfrac{33}{36}$
6.	$\dfrac{4}{16}$	21.	$\dfrac{27}{6}$
7.	$\dfrac{15}{20}$	22.	$\dfrac{40}{18}$
8.	$\dfrac{5}{35}$	23.	$\dfrac{55}{15}$
9.	$\dfrac{12}{42}$	24.	$\dfrac{34}{24}$
10.	$\dfrac{6}{72}$	25.	$\dfrac{56}{22}$
11.	$\dfrac{12}{28}$	26.	$\dfrac{42}{15}$
12.	$\dfrac{14}{77}$	27.	$\dfrac{88}{28}$
13.	$\dfrac{21}{35}$	28.	$2\dfrac{36}{8}$
14.	$\dfrac{15}{40}$	29.	$\dfrac{70}{55}$
15.	$\dfrac{40}{64}$	30.	$1\dfrac{87}{36}$

505 A   Simplify.   Second Half

1.	$\dfrac{2}{4}$		16.	$\dfrac{5}{4}$
2.	$\dfrac{3}{6}$		17.	$\dfrac{26}{10}$
3.	$\dfrac{3}{9}$		18.	$\dfrac{49}{7}$
4.	$\dfrac{6}{9}$		19.	$\dfrac{10}{6}$
5.	$\dfrac{4}{20}$		20.	$\dfrac{44}{48}$
6.	$\dfrac{6}{18}$		21.	$\dfrac{36}{8}$
7.	$\dfrac{12}{16}$		22.	$\dfrac{40}{18}$
8.	$\dfrac{3}{21}$		23.	$\dfrac{26}{6}$
9.	$\dfrac{10}{35}$		24.	$\dfrac{38}{24}$
10.	$\dfrac{5}{60}$		25.	$\dfrac{58}{22}$
11.	$\dfrac{12}{28}$		26.	$\dfrac{42}{15}$
12.	$\dfrac{6}{33}$		27.	$\dfrac{44}{14}$
13.	$\dfrac{15}{40}$		28.	$2\dfrac{54}{12}$
14.	$\dfrac{21}{35}$		29.	$\dfrac{70}{55}$
15.	$\dfrac{20}{32}$		30.	$1\dfrac{75}{36}$

505 B		Simplify.		First Half

1.	$\dfrac{6}{12}$		16.	$\dfrac{10}{8}$
2.	$\dfrac{5}{15}$		17.	$1\dfrac{24}{15}$
3.	$\dfrac{5}{30}$		18.	$2\dfrac{30}{6}$
4.	$\dfrac{6}{30}$		19.	$\dfrac{20}{12}$
5.	$\dfrac{14}{21}$		20.	$\dfrac{66}{72}$
6.	$\dfrac{4}{16}$		21.	$2\dfrac{20}{8}$
7.	$\dfrac{21}{28}$		22.	$\dfrac{60}{27}$
8.	$\dfrac{6}{42}$		23.	$1\dfrac{16}{6}$
9.	$\dfrac{18}{63}$		24.	$\dfrac{51}{36}$
10.	$\dfrac{8}{96}$		25.	$1\dfrac{51}{33}$
11.	$\dfrac{21}{49}$		26.	$\dfrac{56}{20}$
12.	$\dfrac{18}{99}$		27.	$1\dfrac{30}{14}$
13.	$\dfrac{27}{45}$		28.	$2\dfrac{72}{16}$
14.	$\dfrac{21}{56}$		29.	$\dfrac{126}{99}$
15.	$\dfrac{45}{72}$		30.	$1\dfrac{116}{48}$

Simplify.

1.	$\dfrac{3}{6}$		16.	$\dfrac{10}{8}$
2.	$\dfrac{5}{10}$		17.	$\dfrac{39}{15}$
3.	$\dfrac{5}{15}$		18.	$2\dfrac{35}{7}$
4.	$\dfrac{14}{21}$		19.	$\dfrac{20}{12}$
5.	$\dfrac{6}{30}$		20.	$\dfrac{77}{84}$
6.	$\dfrac{7}{21}$		21.	$3\dfrac{24}{16}$
7.	$\dfrac{18}{24}$		22.	$\dfrac{80}{36}$
8.	$\dfrac{5}{35}$		23.	$\dfrac{39}{9}$
9.	$\dfrac{12}{42}$		24.	$\dfrac{38}{24}$
10.	$\dfrac{6}{72}$		25.	$1\dfrac{36}{22}$
11.	$\dfrac{21}{49}$		26.	$\dfrac{56}{20}$
12.	$\dfrac{14}{77}$		27.	$1\dfrac{45}{21}$
13.	$\dfrac{21}{56}$		28.	$2\dfrac{36}{8}$
14.	$\dfrac{27}{45}$		29.	$\dfrac{126}{99}$
15.	$\dfrac{40}{64}$		30.	$1\dfrac{100}{48}$

Math Sprints 5

1.	$\dfrac{1}{2}$	16.	$1\dfrac{1}{4}$
2.	$\dfrac{1}{3}$	17.	$2\dfrac{3}{5}$
3.	$\dfrac{1}{6}$	18.	7
4.	$\dfrac{1}{5}$	19.	$1\dfrac{2}{3}$
5.	$\dfrac{2}{3}$	20.	$\dfrac{11}{12}$
6.	$\dfrac{1}{4}$	21.	$4\dfrac{1}{2}$
7.	$\dfrac{3}{4}$	22.	$2\dfrac{2}{9}$
8.	$\dfrac{1}{7}$	23.	$3\dfrac{2}{3}$
9.	$\dfrac{2}{7}$	24.	$1\dfrac{5}{12}$
10.	$\dfrac{1}{12}$	25.	$2\dfrac{6}{11}$
11.	$\dfrac{3}{7}$	26.	$2\dfrac{4}{5}$
12.	$\dfrac{2}{11}$	27.	$3\dfrac{1}{7}$
13.	$\dfrac{3}{5}$	28.	$6\dfrac{1}{2}$
14.	$\dfrac{3}{8}$	29.	$1\dfrac{3}{11}$
15.	$\dfrac{5}{8}$	30.	$3\dfrac{5}{12}$

Math Sprints 5

1.	$\dfrac{1}{2}$	16.	$1\dfrac{1}{4}$
2.	$\dfrac{1}{2}$	17.	$2\dfrac{3}{5}$
3.	$\dfrac{1}{3}$	18.	7
4.	$\dfrac{2}{3}$	19.	$1\dfrac{2}{3}$
5.	$\dfrac{1}{5}$	20.	$\dfrac{11}{12}$
6.	$\dfrac{1}{3}$	21.	$4\dfrac{1}{2}$
7.	$\dfrac{3}{4}$	22.	$2\dfrac{2}{9}$
8.	$\dfrac{1}{7}$	23.	$4\dfrac{1}{3}$
9.	$\dfrac{2}{7}$	24.	$1\dfrac{7}{12}$
10.	$\dfrac{1}{12}$	25.	$2\dfrac{7}{11}$
11.	$\dfrac{3}{7}$	26.	$2\dfrac{4}{5}$
12.	$\dfrac{2}{11}$	27.	$3\dfrac{1}{7}$
13.	$\dfrac{3}{8}$	28.	$6\dfrac{1}{2}$
14.	$\dfrac{3}{5}$	29.	$1\dfrac{3}{11}$
15.	$\dfrac{5}{8}$	30.	$3\dfrac{1}{12}$

Simplify.

1.	$\dfrac{8}{8}$		16.	$\dfrac{9}{24}$
2.	$\dfrac{6}{12}$		17.	$\dfrac{10}{15}$
3.	$\dfrac{3}{12}$		18.	$\dfrac{12}{20}$
4.	$\dfrac{15}{20}$		19.	$\dfrac{8}{28}$
5.	$\dfrac{3}{15}$		20.	$\dfrac{42}{24}$
6.	$\dfrac{6}{18}$		21.	$\dfrac{24}{20}$
7.	$\dfrac{15}{15}$		22.	$\dfrac{36}{24}$
8.	$\dfrac{12}{18}$		23.	$\dfrac{30}{72}$
9.	$\dfrac{15}{50}$		24.	$\dfrac{40}{64}$
10.	$\dfrac{8}{20}$		25.	$\dfrac{28}{48}$
11.	$\dfrac{4}{18}$		26.	$\dfrac{36}{63}$
12.	$\dfrac{40}{18}$		27.	$\dfrac{60}{108}$
13.	$\dfrac{12}{28}$		28.	$\dfrac{144}{81}$
14.	$\dfrac{22}{12}$		29.	$\dfrac{198}{81}$
15.	$\dfrac{5}{60}$		30.	$\dfrac{248}{96}$

1.	$\dfrac{4}{4}$		16.	$\dfrac{12}{32}$
2.	$\dfrac{3}{6}$		17.	$\dfrac{8}{12}$
3.	$\dfrac{2}{8}$		18.	$\dfrac{18}{30}$
4.	$\dfrac{12}{16}$		19.	$\dfrac{12}{42}$
5.	$\dfrac{4}{20}$		20.	$\dfrac{49}{28}$
6.	$\dfrac{6}{18}$		21.	$\dfrac{30}{25}$
7.	$\dfrac{13}{13}$		22.	$\dfrac{36}{24}$
8.	$\dfrac{10}{15}$		23.	$\dfrac{30}{72}$
9.	$\dfrac{12}{40}$		24.	$\dfrac{21}{8}$
10.	$\dfrac{6}{15}$		25.	$\dfrac{21}{36}$
11.	$\dfrac{6}{27}$		26.	$\dfrac{100}{28}$
12.	$\dfrac{60}{27}$		27.	$\dfrac{46}{18}$
13.	$\dfrac{12}{28}$		28.	$\dfrac{62}{24}$
14.	$\dfrac{33}{18}$		29.	$\dfrac{144}{81}$
15.	$\dfrac{4}{48}$		30.	$\dfrac{198}{81}$

1.	$\dfrac{12}{12}$	16.	$\dfrac{24}{64}$
2.	$\dfrac{15}{30}$	17.	$\dfrac{14}{21}$
3.	$\dfrac{6}{24}$	18.	$\dfrac{24}{40}$
4.	$\dfrac{18}{24}$	19.	$\dfrac{16}{56}$
5.	$\dfrac{7}{35}$	20.	$\dfrac{56}{32}$
6.	$\dfrac{9}{27}$	21.	$\dfrac{36}{30}$
7.	$\dfrac{53}{53}$	22.	$\dfrac{45}{30}$
8.	$\dfrac{22}{33}$	23.	$\dfrac{35}{84}$
9.	$\dfrac{21}{70}$	24.	$\dfrac{45}{72}$
10.	$\dfrac{16}{40}$	25.	$\dfrac{42}{72}$
11.	$\dfrac{18}{81}$	26.	$\dfrac{52}{91}$
12.	$\dfrac{60}{27}$	27.	$\dfrac{65}{117}$
13.	$\dfrac{18}{42}$	28.	$\dfrac{208}{117}$
14.	$\dfrac{66}{36}$	29.	$\dfrac{286}{117}$
15.	$\dfrac{7}{84}$	30.	$\dfrac{279}{108}$

506 B | Simplify. | Second Half

1.	$\dfrac{9}{9}$		16.	$\dfrac{18}{48}$
2.	$\dfrac{10}{20}$		17.	$\dfrac{18}{27}$
3.	$\dfrac{3}{12}$		18.	$\dfrac{21}{35}$
4.	$\dfrac{18}{24}$		19.	$\dfrac{16}{56}$
5.	$\dfrac{8}{40}$		20.	$\dfrac{63}{36}$
6.	$\dfrac{7}{21}$		21.	$\dfrac{42}{35}$
7.	$\dfrac{42}{42}$		22.	$\dfrac{45}{30}$
8.	$\dfrac{12}{18}$		23.	$\dfrac{35}{84}$
9.	$\dfrac{18}{60}$		24.	$\dfrac{42}{16}$
10.	$\dfrac{18}{45}$		25.	$\dfrac{49}{84}$
11.	$\dfrac{12}{54}$		26.	$\dfrac{125}{35}$
12.	$\dfrac{120}{54}$		27.	$\dfrac{69}{27}$
13.	$\dfrac{21}{49}$		28.	$\dfrac{124}{48}$
14.	$\dfrac{55}{30}$		29.	$\dfrac{208}{117}$
15.	$\dfrac{7}{84}$		30.	$\dfrac{286}{117}$

506 A & B Answer Sheet First Half

1.	1	16.	$\dfrac{3}{8}$
2.	$\dfrac{1}{2}$	17.	$\dfrac{2}{3}$
3.	$\dfrac{1}{4}$	18.	$\dfrac{3}{5}$
4.	$\dfrac{3}{4}$	19.	$\dfrac{2}{7}$
5.	$\dfrac{1}{5}$	20.	$1\dfrac{3}{4}$
6.	$\dfrac{1}{3}$	21.	$1\dfrac{1}{5}$
7.	1	22.	$1\dfrac{1}{2}$
8.	$\dfrac{2}{3}$	23.	$\dfrac{5}{12}$
9.	$\dfrac{3}{10}$	24.	$\dfrac{5}{8}$
10.	$\dfrac{2}{5}$	25.	$\dfrac{7}{12}$
11.	$\dfrac{2}{9}$	26.	$\dfrac{4}{7}$
12.	$2\dfrac{2}{9}$	27.	$\dfrac{5}{9}$
13.	$\dfrac{3}{7}$	28.	$1\dfrac{7}{9}$
14.	$1\dfrac{5}{6}$	29.	$2\dfrac{4}{9}$
15.	$\dfrac{1}{12}$	30.	$2\dfrac{7}{12}$

	506 A & B		Answer Sheet		Second Half

1.	1	16.	$\dfrac{3}{8}$
2.	$\dfrac{1}{2}$	17.	$\dfrac{2}{3}$
3.	$\dfrac{1}{4}$	18.	$\dfrac{3}{5}$
4.	$\dfrac{3}{4}$	19.	$\dfrac{2}{7}$
5.	$\dfrac{1}{5}$	20.	$1\dfrac{3}{4}$
6.	$\dfrac{1}{3}$	21.	$1\dfrac{1}{5}$
7.	1	22.	$1\dfrac{1}{2}$
8.	$\dfrac{2}{3}$	23.	$\dfrac{5}{12}$
9.	$\dfrac{3}{10}$	24.	$2\dfrac{5}{8}$
10.	$\dfrac{2}{5}$	25.	$\dfrac{7}{12}$
11.	$\dfrac{2}{9}$	26.	$3\dfrac{4}{7}$
12.	$2\dfrac{2}{9}$	27.	$2\dfrac{5}{9}$
13.	$\dfrac{3}{7}$	28.	$2\dfrac{7}{12}$
14.	$1\dfrac{5}{6}$	29.	$1\dfrac{7}{9}$
15.	$\dfrac{1}{12}$	30.	$2\dfrac{4}{9}$

Math Sprints 5

Write the missing number.

1.	$\dfrac{1}{2} = \dfrac{}{4}$	13.	$\dfrac{3}{8} = \dfrac{18}{}$
2.	$\dfrac{1}{4} = \dfrac{}{8}$	14.	$\dfrac{4}{6} = \dfrac{8}{}$
3.	$\dfrac{1}{3} = \dfrac{}{9}$	15.	$\dfrac{8}{9} = \dfrac{88}{}$
4.	$\dfrac{1}{5} = \dfrac{}{20}$	16.	$\dfrac{3}{7} = \dfrac{}{63}$
5.	$\dfrac{1}{2} = \dfrac{}{18}$	17.	$\dfrac{1}{8} = \dfrac{12}{}$
6.	$\dfrac{3}{10} = \dfrac{9}{}$	18.	$\dfrac{1}{4} = \dfrac{}{32}$
7.	$\dfrac{4}{8} = \dfrac{}{16}$	19.	$\dfrac{1}{2} = \dfrac{}{100}$
8.	$\dfrac{1}{2} = \dfrac{}{20}$	20.	$\dfrac{5}{11} = \dfrac{}{77}$
9.	$\dfrac{5}{7} = \dfrac{15}{}$	21.	$\dfrac{1}{7} = \dfrac{4}{}$
10.	$\dfrac{1}{8} = \dfrac{}{32}$	22.	$\dfrac{4}{9} = \dfrac{36}{}$
11.	$\dfrac{2}{3} = \dfrac{}{9}$	23.	$\dfrac{9}{10} = \dfrac{}{100}$
12.	$\dfrac{4}{5} = \dfrac{16}{}$	24.	$\dfrac{3}{16} = \dfrac{9}{}$

Write the missing number.

1.	$\dfrac{1}{2} = \dfrac{}{4}$	13.	$\dfrac{3}{8} = \dfrac{}{40}$
2.	$\dfrac{1}{3} = \dfrac{}{6}$	14.	$\dfrac{9}{10} = \dfrac{63}{}$
3.	$\dfrac{1}{3} = \dfrac{}{9}$	15.	$\dfrac{5}{8} = \dfrac{}{48}$
4.	$\dfrac{1}{} = \dfrac{3}{12}$	16.	$\dfrac{3}{7} = \dfrac{}{63}$
5.	$\dfrac{1}{2} = \dfrac{6}{}$	17.	$\dfrac{1}{8} = \dfrac{12}{}$
6.	$\dfrac{2}{5} = \dfrac{4}{}$	18.	$\dfrac{1}{4} = \dfrac{}{32}$
7.	$\dfrac{4}{8} = \dfrac{}{16}$	19.	$\dfrac{1}{2} = \dfrac{}{100}$
8.	$\dfrac{}{4} = \dfrac{9}{12}$	20.	$\dfrac{4}{7} = \dfrac{20}{}$
9.	$\dfrac{3}{4} = \dfrac{21}{}$	21.	$\dfrac{6}{7} = \dfrac{}{28}$
10.	$\dfrac{3}{8} = \dfrac{}{32}$	22.	$\dfrac{4}{9} = \dfrac{36}{}$
11.	$\dfrac{1}{} = \dfrac{7}{42}$	23.	$\dfrac{5}{} = \dfrac{45}{108}$
12.	$\dfrac{4}{5} = \dfrac{16}{}$	24.	$\dfrac{7}{13} = \dfrac{28}{}$

1.	$\dfrac{1}{2} = \dfrac{}{4}$	13.	$\dfrac{1}{2} = \dfrac{}{96}$
2.	$\dfrac{1}{4} = \dfrac{}{8}$	14.	$\dfrac{2}{24} = \dfrac{1}{}$
3.	$\dfrac{1}{} = \dfrac{3}{9}$	15.	$\dfrac{8}{9} = \dfrac{88}{}$
4.	$\dfrac{}{24} = \dfrac{1}{6}$	16.	$\dfrac{3}{8} = \dfrac{}{72}$
5.	$\dfrac{1}{3} = \dfrac{}{27}$	17.	$\dfrac{8}{9} = \dfrac{}{108}$
6.	$\dfrac{5}{} = \dfrac{1}{6}$	18.	$\dfrac{1}{20} = \dfrac{}{160}$
7.	$\dfrac{16}{24} = \dfrac{}{12}$	19.	$\dfrac{1}{} = \dfrac{9}{450}$
8.	$\dfrac{21}{30} = \dfrac{7}{}$	20.	$\dfrac{5}{7} = \dfrac{}{49}$
9.	$\dfrac{2}{3} = \dfrac{14}{}$	21.	$\dfrac{7}{12} = \dfrac{}{48}$
10.	$\dfrac{1}{12} = \dfrac{}{48}$	22.	$\dfrac{9}{12} = \dfrac{}{108}$
11.	$\dfrac{}{11} = \dfrac{12}{22}$	23.	$\dfrac{1}{5} = \dfrac{18}{}$
12.	$\dfrac{5}{8} = \dfrac{}{32}$	24.	$\dfrac{36}{} = \dfrac{3}{4}$

1.	$\dfrac{1}{2} = \dfrac{}{4}$	13.	$\dfrac{1}{2} = \dfrac{}{30}$
2.	$\dfrac{2}{4} = \dfrac{1}{}$	14.	$\dfrac{9}{10} = \dfrac{63}{}$
3.	$\dfrac{1}{} = \dfrac{3}{9}$	15.	$\dfrac{5}{9} = \dfrac{}{54}$
4.	$\dfrac{1}{5} = \dfrac{}{20}$	16.	$\dfrac{3}{8} = \dfrac{}{72}$
5.	$\dfrac{3}{4} = \dfrac{9}{}$	17.	$\dfrac{8}{9} = \dfrac{}{108}$
6.	$\dfrac{2}{20} = \dfrac{1}{}$	18.	$\dfrac{1}{20} = \dfrac{}{160}$
7.	$\dfrac{16}{24} = \dfrac{}{12}$	19.	$\dfrac{1}{} = \dfrac{9}{450}$
8.	$\dfrac{2}{} = \dfrac{12}{18}$	20.	$\dfrac{6}{7} = \dfrac{30}{}$
9.	$\dfrac{5}{7} = \dfrac{20}{}$	21.	$\dfrac{3}{12} = \dfrac{}{96}$
10.	$\dfrac{3}{12} = \dfrac{}{48}$	22.	$\dfrac{9}{12} = \dfrac{}{108}$
11.	$\dfrac{}{12} = \dfrac{12}{24}$	23.	$\dfrac{5}{} = \dfrac{45}{108}$
12.	$\dfrac{5}{8} = \dfrac{}{32}$	24.	$\dfrac{9}{13} = \dfrac{36}{}$

507 A & B   Answer Sheet   First Half

1.	2	13.	48
2.	2	14.	12
3.	3	15.	99
4.	4	16.	27
5.	9	17.	96
6.	30	18.	8
7.	8	19.	50
8.	10	20.	35
9.	21	21.	28
10.	4	22.	81
11.	6	23.	90
12.	20	24.	48

1.	2	13.	15
2.	2	14.	70
3.	3	15.	30
4.	4	16.	27
5.	12	17.	96
6.	10	18.	8
7.	8	19.	50
8.	3	20.	35
9.	28	21.	24
10.	12	22.	81
11.	6	23.	12
12.	20	24.	52

Math Sprints 5

1.	$\dfrac{1}{9} + \dfrac{4}{9} =$	11.	$\dfrac{1}{2} + \dfrac{1}{8} =$
2.	$\dfrac{1}{5} + \dfrac{2}{5} =$	12.	$\dfrac{1}{7} + \dfrac{3}{14} =$
3.	$\dfrac{1}{6} + \dfrac{1}{3} =$	13.	$\dfrac{2}{9} + \dfrac{1}{3} =$
4.	$\dfrac{3}{4} + \dfrac{1}{4} =$	14.	$\dfrac{2}{5} + \dfrac{6}{10} =$
5.	$\dfrac{1}{8} + \dfrac{1}{4} =$	15.	$\dfrac{2}{16} + \dfrac{3}{8} =$
6.	$\dfrac{1}{8} + \dfrac{1}{8} =$	16.	$\dfrac{3}{10} + \dfrac{9}{30} =$
7.	$\dfrac{1}{3} + \dfrac{2}{3} =$	17.	$\dfrac{3}{8} + \dfrac{1}{16} =$
8.	$\dfrac{14}{16} + \dfrac{1}{8} =$	18.	$\dfrac{5}{12} + \dfrac{10}{24} =$
9.	$\dfrac{1}{5} + \dfrac{1}{10} =$	19.	$\dfrac{1}{8} + \dfrac{2}{16} =$
10.	$\dfrac{2}{14} + \dfrac{5}{7} =$	20.	$\dfrac{1}{12} + \dfrac{3}{36} =$

508 A Add. Answer should be in simplest form. Second Half

1.	$\dfrac{1}{2}+\dfrac{1}{2}=$		11.	$\dfrac{2}{4}+\dfrac{1}{8}=$
2.	$\dfrac{1}{4}+\dfrac{2}{4}=$		12.	$\dfrac{2}{3}+\dfrac{3}{12}=$
3.	$\dfrac{1}{3}+\dfrac{1}{3}=$		13.	$\dfrac{3}{5}+\dfrac{2}{10}=$
4.	$\dfrac{3}{4}+\dfrac{1}{4}=$		14.	$\dfrac{2}{5}+\dfrac{6}{10}=$
5.	$\dfrac{7}{16}+\dfrac{3}{16}=$		15.	$\dfrac{1}{18}+\dfrac{4}{9}=$
6.	$\dfrac{1}{8}+\dfrac{1}{4}=$		16.	$\dfrac{3}{10}+\dfrac{9}{30}=$
7.	$\dfrac{1}{3}+\dfrac{1}{9}=$		17.	$\dfrac{3}{7}+\dfrac{5}{21}=$
8.	$\dfrac{14}{16}+\dfrac{1}{8}=$		18.	$\dfrac{5}{12}+\dfrac{10}{24}=$
9.	$\dfrac{2}{3}+\dfrac{2}{3}=$		19.	$\dfrac{1}{8}+\dfrac{2}{16}=$
10.	$\dfrac{4}{14}+\dfrac{4}{7}=$		20.	$\dfrac{1}{12}+\dfrac{3}{36}=$

508 B   Add. Answer should be in simplest form.   First Half

1.	$\dfrac{1}{9} + \dfrac{4}{9} =$	11.	$\dfrac{3}{16} + \dfrac{3}{16} + \dfrac{1}{4} =$
2.	$\dfrac{1}{5} + \dfrac{2}{5} =$	12.	$\dfrac{2}{7} + \dfrac{1}{14} =$
3.	$\dfrac{1}{7} + \dfrac{5}{14} =$	13.	$\dfrac{4}{18} + \dfrac{3}{9} =$
4.	$\dfrac{4}{5} + \dfrac{1}{10} + \dfrac{1}{10} =$	14.	$\dfrac{1}{2} + \dfrac{1}{6} + \dfrac{1}{3} =$
5.	$\dfrac{3}{16} + \dfrac{3}{16} =$	15.	$\dfrac{1}{9} + \dfrac{7}{18} =$
6.	$\dfrac{3}{16} + \dfrac{1}{16} =$	16.	$\dfrac{2}{5} + \dfrac{1}{10} + \dfrac{1}{10} =$
7.	$\dfrac{10}{24} + \dfrac{7}{12} =$	17.	$\dfrac{1}{4} + \dfrac{1}{8} + \dfrac{1}{16} =$
8.	$\dfrac{7}{8} + \dfrac{1}{16} + \dfrac{1}{16} =$	18.	$\dfrac{5}{12} + \dfrac{2}{24} + \dfrac{1}{3} =$
9.	$\dfrac{1}{5} + \dfrac{1}{10} =$	19.	$\dfrac{1}{12} + \dfrac{2}{24} + \dfrac{3}{36} =$
10.	$\dfrac{2}{7} + \dfrac{8}{14} =$	20.	$\dfrac{1}{12} + \dfrac{1}{24} + \dfrac{2}{48} =$

508 B Add. Answer should be in simplest form. **Second Half**

1.	$\dfrac{1}{2} + \dfrac{1}{2} =$	11.	$\dfrac{2}{16} + \dfrac{4}{16} + \dfrac{1}{4} =$
2.	$\dfrac{1}{4} + \dfrac{1}{4} + \dfrac{1}{4} =$	12.	$\dfrac{1}{3} + \dfrac{1}{2} + \dfrac{1}{12} =$
3.	$\dfrac{2}{6} + \dfrac{1}{3} =$	13.	$\dfrac{2}{5} + \dfrac{6}{15} =$
4.	$\dfrac{4}{5} + \dfrac{1}{10} + \dfrac{1}{10} =$	14.	$\dfrac{1}{2} + \dfrac{1}{6} + \dfrac{1}{3} =$
5.	$\dfrac{10}{24} + \dfrac{5}{24} =$	15.	$\dfrac{7}{18} + \dfrac{1}{9} =$
6.	$\dfrac{4}{16} + \dfrac{2}{16} =$	16.	$\dfrac{2}{5} + \dfrac{1}{10} + \dfrac{1}{10} =$
7.	$\dfrac{2}{6} + \dfrac{1}{9} =$	17.	$\dfrac{1}{3} + \dfrac{1}{7} + \dfrac{4}{21} =$
8.	$\dfrac{7}{8} + \dfrac{2}{16} =$	18.	$\dfrac{5}{12} + \dfrac{2}{24} + \dfrac{1}{3} =$
9.	$\dfrac{1}{3} + \dfrac{1}{3} + \dfrac{2}{3} =$	19.	$\dfrac{1}{12} + \dfrac{2}{24} + \dfrac{3}{36} =$
10.	$\dfrac{3}{7} + \dfrac{6}{14} =$	20.	$\dfrac{1}{24} + \dfrac{6}{48} =$

Math Sprints 5

1.	$\dfrac{5}{9}$	11.	$\dfrac{5}{8}$
2.	$\dfrac{3}{5}$	12.	$\dfrac{5}{14}$
3.	$\dfrac{1}{2}$	13.	$\dfrac{5}{9}$
4.	1	14.	1
5.	$\dfrac{3}{8}$	15.	$\dfrac{1}{2}$
6.	$\dfrac{1}{4}$	16.	$\dfrac{3}{5}$
7.	1	17.	$\dfrac{7}{16}$
8.	1	18.	$\dfrac{5}{6}$
9.	$\dfrac{3}{10}$	19.	$\dfrac{1}{4}$
10.	$\dfrac{6}{7}$	20.	$\dfrac{1}{6}$

508 A & B   Answer Sheet   Second Half

1.	1	11.	$\dfrac{5}{8}$
2.	$\dfrac{3}{4}$	12.	$\dfrac{11}{12}$
3.	$\dfrac{2}{3}$	13.	$\dfrac{4}{5}$
4.	1	14.	1
5.	$\dfrac{5}{8}$	15.	$\dfrac{1}{2}$
6.	$\dfrac{3}{8}$	16.	$\dfrac{3}{5}$
7.	$\dfrac{4}{9}$	17.	$\dfrac{2}{3}$
8.	1	18.	$\dfrac{5}{6}$
9.	$1\dfrac{1}{3}$	19.	$\dfrac{1}{4}$
10.	$\dfrac{6}{7}$	20.	$\dfrac{1}{6}$

Math Sprints 5

Add. Answer should be in simplest form.

1.	$\dfrac{1}{2} + \dfrac{1}{4} =$		13.	$\dfrac{3}{8} + \dfrac{7}{40} =$
2.	$\dfrac{1}{6} + \dfrac{1}{3} =$		14.	$\dfrac{3}{7} + \dfrac{2}{5} =$
3.	$\dfrac{5}{12} + \dfrac{1}{6} =$		15.	$\dfrac{4}{5} + \dfrac{3}{8} =$
4.	$\dfrac{2}{3} + \dfrac{1}{12} =$		16.	$\dfrac{5}{6} + \dfrac{3}{4} =$
5.	$\dfrac{2}{9} + \dfrac{1}{2} =$		17.	$\dfrac{1}{6} + \dfrac{3}{11} =$
6.	$\dfrac{2}{5} + \dfrac{3}{10} =$		18.	$\dfrac{3}{10} + \dfrac{3}{8} =$
7.	$\dfrac{3}{8} + \dfrac{3}{4} =$		19.	$\dfrac{3}{10} + \dfrac{5}{12} =$
8.	$\dfrac{2}{3} + \dfrac{3}{5} =$		20.	$\dfrac{2}{3} + \dfrac{7}{10} =$
9.	$\dfrac{2}{7} + \dfrac{1}{3} =$		21.	$\dfrac{5}{6} + \dfrac{7}{9} =$
10.	$\dfrac{5}{6} + \dfrac{1}{10} =$		22.	$\dfrac{5}{12} + \dfrac{1}{8} =$
11.	$\dfrac{1}{4} + \dfrac{1}{6} =$		23.	$\dfrac{3}{8} + \dfrac{5}{12} =$
12.	$\dfrac{5}{9} + \dfrac{5}{6} =$		24.	$\dfrac{1}{6} + \dfrac{3}{10} =$

Math Sprints 5

Add. Answer should be in simplest form.   Second Half

1.	$\dfrac{1}{2} + \dfrac{1}{4} =$	13.	$\dfrac{7}{20} + \dfrac{1}{5} =$
2.	$\dfrac{2}{3} + \dfrac{1}{6} =$	14.	$\dfrac{10}{12} + \dfrac{1}{3} =$
3.	$\dfrac{5}{12} + \dfrac{1}{6} =$	15.	$\dfrac{4}{5} + \dfrac{3}{8} =$
4.	$\dfrac{1}{3} + \dfrac{2}{9} =$	16.	$\dfrac{5}{6} + \dfrac{1}{4} =$
5.	$\dfrac{2}{9} + \dfrac{1}{2} =$	17.	$\dfrac{1}{6} + \dfrac{3}{11} =$
6.	$\dfrac{3}{5} + \dfrac{1}{10} =$	18.	$\dfrac{3}{10} + \dfrac{2}{8} =$
7.	$\dfrac{3}{8} + \dfrac{3}{4} =$	19.	$\dfrac{3}{10} + \dfrac{5}{12} =$
8.	$\dfrac{1}{3} + \dfrac{2}{5} =$	20.	$\dfrac{1}{3} + \dfrac{9}{10} =$
9.	$\dfrac{2}{7} + \dfrac{1}{3} =$	21.	$\dfrac{5}{6} + \dfrac{7}{9} =$
10.	$\dfrac{3}{5} + \dfrac{1}{2} =$	22.	$1\dfrac{2}{6} + 1\dfrac{4}{9} =$
11.	$\dfrac{1}{4} + \dfrac{1}{6} =$	23.	$\dfrac{3}{8} + \dfrac{5}{12} =$
12.	$\dfrac{4}{9} + \dfrac{1}{6} =$	24.	$\dfrac{1}{6} + \dfrac{3}{10} =$

509 B		Subtract. Answer should be in simplest form.		First Half

1.	$1\dfrac{1}{2} - \dfrac{3}{4} =$	13.	$\dfrac{4}{5} - \dfrac{1}{4} =$
2.	$\dfrac{4}{5} - \dfrac{3}{10} =$	14.	$1\dfrac{3}{7} - \dfrac{3}{5} =$
3.	$1\dfrac{1}{6} - \dfrac{7}{12} =$	15.	$1\dfrac{5}{8} - \dfrac{9}{20} =$
4.	$1\dfrac{1}{3} - \dfrac{7}{12} =$	16.	$2\dfrac{1}{4} - \dfrac{2}{3} =$
5.	$\dfrac{8}{9} - \dfrac{3}{18} =$	17.	$\dfrac{5}{6} - \dfrac{13}{33} =$
6.	$1\dfrac{3}{5} - \dfrac{9}{10} =$	18.	$\dfrac{7}{8} - \dfrac{1}{5} =$
7.	$1\dfrac{3}{4} - \dfrac{5}{8} =$	19.	$1\dfrac{3}{10} - \dfrac{7}{12} =$
8.	$1\dfrac{2}{3} - \dfrac{2}{5} =$	20.	$2\dfrac{1}{10} - \dfrac{11}{15} =$
9.	$1\dfrac{1}{3} - \dfrac{5}{7} =$	21.	$1\dfrac{5}{6} - \dfrac{2}{9} =$
10.	$1\dfrac{1}{6} - \dfrac{7}{30} =$	22.	$\dfrac{7}{8} - \dfrac{1}{3} =$
11.	$1\dfrac{3}{8} - \dfrac{23}{24} =$	23.	$\dfrac{7}{8} - \dfrac{1}{12} =$
12.	$1\dfrac{8}{9} - \dfrac{1}{2} =$	24.	$1\dfrac{9}{30} - \dfrac{5}{6} =$

509 B Subtract. Answer should be in simplest form. Second Half

1.	$1\dfrac{1}{2} - \dfrac{3}{4} =$	13.	$\dfrac{4}{5} - \dfrac{1}{4} =$
2.	$1 - \dfrac{2}{12} =$	14.	$1\dfrac{11}{12} - \dfrac{3}{4} =$
3.	$1\dfrac{1}{6} - \dfrac{7}{12} =$	15.	$1\dfrac{5}{8} - \dfrac{9}{20} =$
4.	$1\dfrac{1}{3} - \dfrac{7}{9} =$	16.	$1\dfrac{1}{4} - \dfrac{1}{6} =$
5.	$\dfrac{8}{9} - \dfrac{3}{18} =$	17.	$\dfrac{5}{6} - \dfrac{13}{33} =$
6.	$1\dfrac{2}{5} - \dfrac{7}{10} =$	18.	$1\dfrac{3}{10} - \dfrac{3}{4} =$
7.	$1\dfrac{3}{4} - \dfrac{5}{8} =$	19.	$1\dfrac{3}{10} - \dfrac{7}{12} =$
8.	$1\dfrac{1}{3} - \dfrac{3}{5} =$	20.	$2\dfrac{1}{30} - \dfrac{4}{5} =$
9.	$1\dfrac{1}{3} - \dfrac{5}{7} =$	21.	$1\dfrac{5}{6} - \dfrac{2}{9} =$
10.	$1\dfrac{1}{2} - \dfrac{2}{5} =$	22.	$3\dfrac{4}{9} - \dfrac{2}{3} =$
11.	$1\dfrac{3}{8} - \dfrac{23}{24} =$	23.	$\dfrac{7}{8} - \dfrac{1}{12} =$
12.	$1\dfrac{1}{9} - \dfrac{1}{2} =$	24.	$1\dfrac{9}{30} - \dfrac{5}{6} =$

509 A & B   Answer Sheet   First Half

1.	$\dfrac{3}{4}$	13.	$\dfrac{11}{20}$
2.	$\dfrac{1}{2}$	14.	$\dfrac{29}{35}$
3.	$\dfrac{7}{12}$	15.	$1\dfrac{7}{40}$
4.	$\dfrac{3}{4}$	16.	$1\dfrac{7}{12}$
5.	$\dfrac{13}{18}$	17.	$\dfrac{29}{66}$
6.	$\dfrac{7}{10}$	18.	$\dfrac{27}{40}$
7.	$1\dfrac{1}{8}$	19.	$\dfrac{43}{60}$
8.	$1\dfrac{4}{15}$	20.	$1\dfrac{11}{30}$
9.	$\dfrac{13}{21}$	21.	$1\dfrac{11}{18}$
10.	$\dfrac{14}{15}$	22.	$\dfrac{13}{24}$
11.	$\dfrac{5}{12}$	23.	$\dfrac{19}{24}$
12.	$1\dfrac{7}{18}$	24.	$\dfrac{7}{15}$

1.	$\dfrac{3}{4}$	13.	$\dfrac{11}{20}$
2.	$\dfrac{5}{6}$	14.	$1\dfrac{1}{6}$
3.	$\dfrac{7}{12}$	15.	$1\dfrac{7}{40}$
4.	$\dfrac{5}{9}$	16.	$1\dfrac{1}{12}$
5.	$\dfrac{13}{18}$	17.	$\dfrac{29}{66}$
6.	$\dfrac{7}{10}$	18.	$\dfrac{11}{20}$
7.	$1\dfrac{1}{8}$	19.	$\dfrac{43}{60}$
8.	$\dfrac{11}{15}$	20.	$1\dfrac{7}{30}$
9.	$\dfrac{13}{21}$	21.	$1\dfrac{11}{18}$
10.	$1\dfrac{1}{10}$	22.	$2\dfrac{7}{9}$
11.	$\dfrac{5}{12}$	23.	$\dfrac{19}{24}$
12.	$\dfrac{11}{18}$	24.	$\dfrac{7}{15}$

510 A Multiply. Answer should be in simplest form. First Half

1.	$\dfrac{1}{2} \times 2 =$	16.	$8 \times \dfrac{3}{4} =$
2.	$\dfrac{1}{4} \times 2 =$	17.	$\dfrac{3}{8} \times 4 =$
3.	$\dfrac{1}{2} \times 3 =$	18.	$\dfrac{2}{7} \times 6 =$
4.	$\dfrac{1}{5} \times 4 =$	19.	$9 \times \dfrac{1}{6} =$
5.	$\dfrac{1}{8} \times 3 =$	20.	$\dfrac{3}{10} \times 7 =$
6.	$\dfrac{1}{12} \times 4 =$	21.	$8 \times \dfrac{4}{9} =$
7.	$\dfrac{1}{11} \times 22 =$	22.	$7 \times \dfrac{5}{6} =$
8.	$\dfrac{2}{3} \times 4 =$	23.	$\dfrac{6}{7} \times 5 =$
9.	$2 \times \dfrac{2}{5} =$	24.	$12 \times \dfrac{8}{9} =$
10.	$3 \times \dfrac{3}{4} =$	25.	$\dfrac{7}{10} \times 8 =$
11.	$\dfrac{1}{5} \times 6 =$	26.	$8 \times \dfrac{2}{9} =$
12.	$4 \times \dfrac{3}{5} =$	27.	$\dfrac{3}{8} \times 9 =$
13.	$5 \times \dfrac{3}{7} =$	28.	$12 \times \dfrac{5}{9} =$
14.	$\dfrac{3}{5} \times 6 =$	29.	$1\dfrac{2}{3} \times 5 =$
15.	$3 \times \dfrac{2}{3} =$	30.	$1\dfrac{2}{7} \times 9 =$

Multiply. Answer should be in simplest form.

1.	$\dfrac{1}{2} \times 2 =$		16.	$10 \times \dfrac{3}{5} =$
2.	$\dfrac{1}{4} \times 3 =$		17.	$\dfrac{3}{8} \times 4 =$
3.	$\dfrac{1}{2} \times 3 =$		18.	$\dfrac{2}{7} \times 6 =$
4.	$\dfrac{1}{5} \times 2 =$		19.	$9 \times \dfrac{2}{6} =$
5.	$\dfrac{1}{8} \times 4 =$		20.	$\dfrac{4}{9} \times 8 =$
6.	$\dfrac{1}{12} \times 3 =$		21.	$7 \times \dfrac{3}{10} =$
7.	$\dfrac{1}{5} \times 10 =$		22.	$7 \times \dfrac{5}{6} =$
8.	$\dfrac{2}{3} \times 4 =$		23.	$\dfrac{5}{7} \times 6 =$
9.	$2 \times \dfrac{3}{5} =$		24.	$12 \times \dfrac{8}{9} =$
10.	$3 \times \dfrac{3}{4} =$		25.	$\dfrac{8}{10} \times 7 =$
11.	$\dfrac{1}{5} \times 6 =$		26.	$8 \times \dfrac{2}{9} =$
12.	$5 \times \dfrac{1}{3} =$		27.	$\dfrac{5}{9} \times 12 =$
13.	$4 \times \dfrac{3}{7} =$		28.	$9 \times \dfrac{3}{8} =$
14.	$\dfrac{3}{5} \times 6 =$		29.	$1\dfrac{2}{3} \times 5 =$
15.	$5 \times \dfrac{2}{5} =$		30.	$1\dfrac{2}{7} \times 9 =$

510 B Multiply. Answer should be in simplest form. First Half

1.	$\dfrac{1}{2} \times 2 =$	16.	$18 \times \dfrac{3}{9} =$
2.	$\dfrac{1}{8} \times 4 =$	17.	$\dfrac{3}{10} \times 5 =$
3.	$\dfrac{1}{4} \times 6 =$	18.	$\dfrac{4}{21} \times 9 =$
4.	$8 \times \dfrac{1}{10} =$	19.	$7 \times \dfrac{3}{14} =$
5.	$\dfrac{1}{8} \times 3 =$	20.	$\dfrac{7}{30} \times 9 =$
6.	$\dfrac{1}{24} \times 8 =$	21.	$4 \times \dfrac{8}{9} =$
7.	$\dfrac{1}{13} \times 26 =$	22.	$10 \times \dfrac{7}{12} =$
8.	$\dfrac{2}{3} \times 4 =$	23.	$\dfrac{10}{21} \times 9 =$
9.	$16 \times \dfrac{1}{20} =$	24.	$11 \times \dfrac{32}{33} =$
10.	$6 \times \dfrac{3}{8} =$	25.	$\dfrac{7}{15} \times 12 =$
11.	$\dfrac{2}{5} \times 3 =$	26.	$12 \times \dfrac{4}{27} =$
12.	$6 \times \dfrac{2}{5} =$	27.	$\dfrac{9}{32} \times 12 =$
13.	$3 \times \dfrac{5}{7} =$	28.	$12 \times \dfrac{10}{18} =$
14.	$\dfrac{9}{10} \times 4 =$	29.	$1\dfrac{7}{18} \times 6 =$
15.	$18 \times \dfrac{2}{18}$	30.	$1\dfrac{11}{70} \times 10 =$

510 B Multiply. Answer should be in simplest form. **Second Half**

1.	$\dfrac{1}{2} \times 2 =$		16.	$12 \times \dfrac{3}{6} =$
2.	$\dfrac{1}{8} \times 6 =$		17.	$\dfrac{3}{10} \times 5 =$
3.	$\dfrac{1}{4} \times 6 =$		18.	$\dfrac{4}{21} \times 9 =$
4.	$4 \times \dfrac{1}{10} =$		19.	$4 \times \dfrac{6}{8} =$
5.	$\dfrac{1}{12} \times 6 =$		20.	$\dfrac{8}{9} \times 4 =$
6.	$\dfrac{1}{24} \times 6 =$		21.	$9 \times \dfrac{7}{30} =$
7.	$\dfrac{1}{14} \times 28 =$		22.	$10 \times \dfrac{7}{12} =$
8.	$\dfrac{2}{3} \times 4 =$		23.	$\dfrac{9}{21} \times 10 =$
9.	$2 \times \dfrac{6}{10} =$		24.	$11 \times \dfrac{32}{33} =$
10.	$6 \times \dfrac{3}{8} =$		25.	$\dfrac{12}{15} \times 7 =$
11.	$\dfrac{2}{5} \times 3 =$		26.	$12 \times \dfrac{4}{27} =$
12.	$5 \times \dfrac{2}{6} =$		27.	$\dfrac{10}{18} \times 12 =$
13.	$3 \times \dfrac{4}{7} =$		28.	$12 \times \dfrac{9}{32} =$
14.	$\dfrac{9}{10} \times 4 =$		29.	$1\dfrac{7}{18} \times 6 =$
15.	$17 \times \dfrac{2}{17} =$		30.	$1\dfrac{11}{70} \times 10 =$

	510 A & B		Answer Sheet		First Half

1.	1	16.	6
2.	$\frac{1}{2}$	17.	$1\frac{1}{2}$
3.	$1\frac{1}{2}$	18.	$1\frac{5}{7}$
4.	$\frac{4}{5}$	19.	$1\frac{1}{2}$
5.	$\frac{3}{8}$	20.	$2\frac{1}{10}$
6.	$\frac{1}{3}$	21.	$3\frac{5}{9}$
7.	2	22.	$5\frac{5}{6}$
8.	$2\frac{2}{3}$	23.	$4\frac{2}{7}$
9.	$\frac{4}{5}$	24.	$10\frac{2}{3}$
10.	$2\frac{1}{4}$	25.	$5\frac{3}{5}$
11.	$1\frac{1}{5}$	26.	$1\frac{7}{9}$
12.	$2\frac{2}{5}$	27.	$3\frac{3}{8}$
13.	$2\frac{1}{7}$	28.	$6\frac{2}{3}$
14.	$3\frac{3}{5}$	29.	$8\frac{1}{3}$
15.	2	30.	$11\frac{4}{7}$

510 A & B Answer Sheet Second Half

#		#	
1.	1	16.	6
2.	$\frac{3}{4}$	17.	$1\frac{1}{2}$
3.	$1\frac{1}{2}$	18.	$1\frac{5}{7}$
4.	$\frac{2}{5}$	19.	3
5.	$\frac{1}{2}$	20.	$3\frac{5}{9}$
6.	$\frac{1}{4}$	21.	$2\frac{1}{10}$
7.	2	22.	$5\frac{5}{6}$
8.	$2\frac{2}{3}$	23.	$4\frac{2}{7}$
9.	$1\frac{1}{5}$	24.	$10\frac{2}{3}$
10.	$2\frac{1}{4}$	25.	$5\frac{3}{5}$
11.	$1\frac{1}{5}$	26.	$1\frac{7}{9}$
12.	$1\frac{2}{3}$	27.	$6\frac{2}{3}$
13.	$1\frac{5}{7}$	28.	$3\frac{3}{8}$
14.	$3\frac{3}{5}$	29.	$8\frac{1}{3}$
15.	2	30.	$11\frac{4}{7}$

| 511 A | | Multiply. Answer should be in simplest form. | | First Half |

1.	$\dfrac{1}{2} \times \dfrac{1}{2} =$		12.	$\dfrac{1}{6} \times \dfrac{5}{6} =$
2.	$\dfrac{1}{2} \times \dfrac{1}{4} =$		13.	$\dfrac{2}{3} \times \dfrac{4}{5} =$
3.	$\dfrac{1}{2} \times \dfrac{3}{4} =$		14.	$\dfrac{1}{2} \times \dfrac{5}{4} =$
4.	$\dfrac{3}{5} \times \dfrac{1}{2} =$		15.	$\dfrac{5}{6} \times \dfrac{1}{5} =$
5.	$\dfrac{4}{5} \times \dfrac{1}{2} =$		16.	$\dfrac{3}{4} \times \dfrac{5}{6} =$
6.	$\dfrac{1}{3} \times \dfrac{1}{4} =$		17.	$\dfrac{5}{8} \times \dfrac{4}{9} =$
7.	$\dfrac{2}{3} \times \dfrac{1}{4} =$		18.	$\dfrac{3}{8} \times \dfrac{16}{17} =$
8.	$\dfrac{2}{3} \times \dfrac{3}{4} =$		19.	$\dfrac{2}{3} \times \dfrac{9}{10} =$
9.	$\dfrac{1}{5} \times \dfrac{1}{3} =$		20.	$\dfrac{5}{6} \times \dfrac{7}{10} =$
10.	$\dfrac{1}{5} \times \dfrac{2}{3} =$		21.	$\dfrac{4}{15} \times \dfrac{8}{4} =$
11.	$\dfrac{3}{5} \times \dfrac{1}{3} =$		22.	$\dfrac{3}{4} \times \dfrac{16}{27} =$

511 A Multiply. Answer should be in simplest form. **Second Half**

1.	$\dfrac{1}{2} \times \dfrac{1}{2} =$		12.	$\dfrac{1}{10} \times \dfrac{5}{6} =$
2.	$\dfrac{1}{2} \times \dfrac{1}{3} =$		13.	$\dfrac{2}{3} \times \dfrac{4}{5} =$
3.	$\dfrac{1}{2} \times \dfrac{2}{4} =$		14.	$\dfrac{1}{2} \times \dfrac{5}{4} =$
4.	$\dfrac{4}{5} \times \dfrac{1}{2} =$		15.	$\dfrac{5}{6} \times \dfrac{2}{5} =$
5.	$\dfrac{4}{5} \times \dfrac{1}{4} =$		16.	$\dfrac{3}{5} \times \dfrac{5}{6} =$
6.	$\dfrac{1}{3} \times \dfrac{1}{4} =$		17.	$\dfrac{3}{4} \times \dfrac{5}{6} =$
7.	$\dfrac{2}{3} \times \dfrac{2}{4} =$		18.	$\dfrac{3}{8} \times \dfrac{16}{21} =$
8.	$\dfrac{2}{3} \times \dfrac{3}{4} =$		19.	$\dfrac{2}{3} \times \dfrac{9}{10} =$
9.	$\dfrac{1}{5} \times \dfrac{1}{4} =$		20.	$\dfrac{5}{6} \times \dfrac{7}{25} =$
10.	$\dfrac{1}{5} \times \dfrac{3}{4} =$		21.	$\dfrac{4}{15} \times \dfrac{8}{4} =$
11.	$\dfrac{3}{5} \times \dfrac{2}{3} =$		22.	$\dfrac{3}{4} \times \dfrac{20}{27} =$

511 B		Multiply. Answer should be in simplest form.		First Half

1.	$\dfrac{2}{4} \times \dfrac{1}{2} =$	12.	$\dfrac{2}{12} \times \dfrac{5}{6} =$
2.	$\dfrac{1}{2} \times \dfrac{2}{8} =$	13.	$\dfrac{4}{6} \times \dfrac{4}{5} =$
3.	$\dfrac{2}{4} \times \dfrac{3}{4} =$	14.	$\dfrac{3}{4} \times \dfrac{5}{6} =$
4.	$\dfrac{6}{10} \times \dfrac{1}{2} =$	15.	$\dfrac{5}{3} \times \dfrac{1}{10} =$
5.	$\dfrac{8}{10} \times \dfrac{1}{2} =$	16.	$\dfrac{3}{8} \times \dfrac{10}{6} =$
6.	$\dfrac{2}{6} \times \dfrac{1}{4} =$	17.	$\dfrac{5}{16} \times \dfrac{8}{9} =$
7.	$\dfrac{4}{6} \times \dfrac{1}{4} =$	18.	$\dfrac{3}{8} \times \dfrac{16}{17} =$
8.	$2 \times \dfrac{1}{3} \times \dfrac{3}{4} =$	19.	$\dfrac{10}{3} \times \dfrac{9}{50} =$
9.	$\dfrac{1}{5} \times \dfrac{1}{3} =$	20.	$\dfrac{5}{12} \times \dfrac{14}{10} =$
10.	$\dfrac{2}{10} \times \dfrac{2}{3} =$	21.	$\dfrac{4}{30} \times \dfrac{8}{2} =$
11.	$3 \times \dfrac{1}{5} \times \dfrac{1}{3} =$	22.	$\dfrac{3}{4} \times \dfrac{16}{27} =$

511 B Multiply. Answer should be in simplest form. **Second Half**

1.	$\dfrac{1}{2} \times \dfrac{1}{2} =$	12.	$\dfrac{2}{20} \times \dfrac{5}{6} =$
2.	$\dfrac{1}{2} \times \dfrac{2}{6} =$	13.	$\dfrac{4}{6} \times \dfrac{4}{5} =$
3.	$\dfrac{2}{4} \times \dfrac{2}{4} =$	14.	$\dfrac{3}{4} \times \dfrac{5}{6} =$
4.	$\dfrac{8}{10} \times \dfrac{1}{2} =$	15.	$\dfrac{5}{3} \times \dfrac{2}{10} =$
5.	$\dfrac{8}{10} \times \dfrac{1}{4} =$	16.	$\dfrac{3}{5} \times \dfrac{5}{6} =$
6.	$\dfrac{2}{6} \times \dfrac{1}{4} =$	17.	$\dfrac{3}{8} \times \dfrac{10}{6} =$
7.	$\dfrac{4}{6} \times \dfrac{2}{4} =$	18.	$\dfrac{3}{8} \times \dfrac{16}{21} =$
8.	$3 \times \dfrac{1}{4} \times \dfrac{2}{3} =$	19.	$\dfrac{10}{3} \times \dfrac{9}{50} =$
9.	$\dfrac{1}{5} \times \dfrac{1}{4} =$	20.	$\dfrac{5}{12} \times \dfrac{14}{25} =$
10.	$\dfrac{2}{10} \times \dfrac{3}{4} =$	21.	$\dfrac{4}{30} \times \dfrac{8}{2} =$
11.	$3 \times \dfrac{1}{5} \times \dfrac{2}{3} =$	22.	$\dfrac{3}{4} \times \dfrac{20}{27} =$

511 A & B	Answer Sheet		First Half

1.	$\dfrac{1}{4}$	12.	$\dfrac{5}{36}$
2.	$\dfrac{1}{8}$	13.	$\dfrac{8}{15}$
3.	$\dfrac{3}{8}$	14.	$\dfrac{5}{8}$
4.	$\dfrac{3}{10}$	15.	$\dfrac{1}{6}$
5.	$\dfrac{2}{5}$	16.	$\dfrac{5}{8}$
6.	$\dfrac{1}{12}$	17.	$\dfrac{5}{18}$
7.	$\dfrac{1}{6}$	18.	$\dfrac{6}{17}$
8.	$\dfrac{1}{2}$	19.	$\dfrac{3}{5}$
9.	$\dfrac{1}{15}$	20.	$\dfrac{7}{12}$
10.	$\dfrac{2}{15}$	21.	$\dfrac{8}{15}$
11.	$\dfrac{1}{5}$	22.	$\dfrac{4}{9}$

| | 511 A & B | | Answer Sheet | | Second Half |

1.	$\dfrac{1}{4}$	12.	$\dfrac{1}{12}$
2.	$\dfrac{1}{6}$	13.	$\dfrac{8}{15}$
3.	$\dfrac{1}{4}$	14.	$\dfrac{5}{8}$
4.	$\dfrac{2}{5}$	15.	$\dfrac{1}{3}$
5.	$\dfrac{1}{5}$	16.	$\dfrac{1}{2}$
6.	$\dfrac{1}{12}$	17.	$\dfrac{5}{8}$
7.	$\dfrac{1}{3}$	18.	$\dfrac{2}{7}$
8.	$\dfrac{1}{2}$	19.	$\dfrac{3}{5}$
9.	$\dfrac{1}{20}$	20.	$\dfrac{7}{30}$
10.	$\dfrac{3}{20}$	21.	$\dfrac{8}{15}$
11.	$\dfrac{2}{5}$	22.	$\dfrac{5}{9}$

512 A   Divide. Answer should be in simplest form.   First Half

1.	$\dfrac{1}{2} \div 2 =$	13.	$\dfrac{2}{3} \div 4 =$
2.	$\dfrac{1}{3} \div 2 =$	14.	$\dfrac{6}{7} \div 3 =$
3.	$\dfrac{1}{4} \div 3 =$	15.	$\dfrac{4}{7} \div 2 =$
4.	$\dfrac{1}{5} \div 2 =$	16.	$\dfrac{4}{9} \div 2 =$
5.	$\dfrac{1}{4} \div 5 =$	17.	$\dfrac{4}{3} \div 8 =$
6.	$\dfrac{1}{7} \div 6 =$	18.	$\dfrac{3}{4} \div 9 =$
7.	$\dfrac{3}{4} \div 3 =$	19.	$\dfrac{4}{8} \div 2 =$
8.	$\dfrac{4}{5} \div 4 =$	20.	$\dfrac{4}{5} \div 10 =$
9.	$\dfrac{5}{7} \div 5 =$	21.	$\dfrac{2}{3} \div 6 =$
10.	$\dfrac{6}{7} \div 6 =$	22.	$\dfrac{5}{12} \div 10 =$
11.	$\dfrac{4}{5} \div 2 =$	23.	$\dfrac{7}{12} \div 21 =$
12.	$\dfrac{1}{4} \div 3 =$	24.	$\dfrac{8}{9} \div 12 =$

512 A Divide. Answer should be in simplest form. Second Half

1.	$\dfrac{1}{2} \div 2 =$		13.	$\dfrac{1}{3} \div 4 =$
2.	$\dfrac{1}{2} \div 3 =$		14.	$\dfrac{6}{7} \div 3 =$
3.	$\dfrac{1}{4} \div 3 =$		15.	$\dfrac{4}{5} \div 2 =$
4.	$\dfrac{1}{3} \div 2 =$		16.	$\dfrac{4}{9} \div 2 =$
5.	$\dfrac{1}{5} \div 4 =$		17.	$\dfrac{4}{5} \div 8 =$
6.	$\dfrac{1}{5} \div 7 =$		18.	$\dfrac{3}{4} \div 9 =$
7.	$\dfrac{3}{7} \div 3 =$		19.	$\dfrac{6}{8} \div 2 =$
8.	$\dfrac{4}{5} \div 2 =$		20.	$\dfrac{6}{5} \div 10 =$
9.	$\dfrac{5}{9} \div 5 =$		21.	$\dfrac{2}{3} \div 6 =$
10.	$\dfrac{6}{5} \div 6 =$		22.	$\dfrac{5}{12} \div 10 =$
11.	$\dfrac{4}{5} \div 2 =$		23.	$\dfrac{7}{12} \div 14 =$
12.	$\dfrac{1}{6} \div 3 =$		24.	$\dfrac{8}{9} \div 12 =$

512 B Divide. Answer should be in simplest form. First Half

1.	$\dfrac{1}{2} \div 2 =$	13.	$\dfrac{4}{3} \div 8 =$
2.	$\dfrac{1}{3} \div 2 =$	14.	$\dfrac{16}{7} \div 8 =$
3.	$\dfrac{1}{4} \div 3 =$	15.	$\dfrac{8}{7} \div 4 =$
4.	$\dfrac{2}{5} \div 4 =$	16.	$\dfrac{8}{9} \div 4 =$
5.	$\dfrac{2}{5} \div 8 =$	17.	$1\dfrac{1}{3} \div 8 =$
6.	$\dfrac{1}{7} \div 6 =$	18.	$\dfrac{5}{4} \div 15 =$
7.	$\dfrac{6}{8} \div 3 =$	19.	$1\dfrac{4}{8} \div 6 =$
8.	$\dfrac{8}{10} \div 4 =$	20.	$1\dfrac{3}{5} \div 20 =$
9.	$\dfrac{5}{7} \div 5 =$	21.	$1\dfrac{1}{3} \div 12 =$
10.	$\dfrac{6}{7} \div 6 =$	22.	$\dfrac{5}{6} \div 20 =$
11.	$\dfrac{8}{10} \div 2 =$	23.	$1\dfrac{1}{6} \div 42 =$
12.	$\dfrac{1}{4} \div 3 =$	24.	$1\dfrac{7}{9} \div 24 =$

512 B Divide. Answer should be in simplest form. **Second Half**

1.	$\dfrac{1}{2} \div 2 =$		13.	$\dfrac{2}{3} \div 8 =$
2.	$\dfrac{1}{2} \div 3 =$		14.	$\dfrac{16}{7} \div 8 =$
3.	$\dfrac{1}{3} \div 4 =$		15.	$\dfrac{8}{5} \div 4 =$
4.	$\dfrac{2}{3} \div 4 =$		16.	$\dfrac{8}{9} \div 4 =$
5.	$\dfrac{2}{5} \div 8 =$		17.	$\dfrac{4}{5} \div 8 =$
6.	$\dfrac{1}{5} \div 7 =$		18.	$\dfrac{5}{4} \div 15 =$
7.	$\dfrac{6}{7} \div 6 =$		19.	$2\dfrac{1}{4} \div 6 =$
8.	$\dfrac{8}{10} \div 2 =$		20.	$1\dfrac{1}{5} \div 10 =$
9.	$\dfrac{5}{9} \div 5 =$		21.	$1\dfrac{1}{3} \div 12 =$
10.	$\dfrac{6}{5} \div 6 =$		22.	$\dfrac{5}{6} \div 20 =$
11.	$\dfrac{8}{10} \div 2 =$		23.	$1\dfrac{1}{6} \div 28 =$
12.	$\dfrac{1}{6} \div 3 =$		24.	$1\dfrac{7}{9} \div 24 =$

Math Sprints 5

1.	$\dfrac{1}{4}$	13.	$\dfrac{1}{6}$
2.	$\dfrac{1}{6}$	14.	$\dfrac{2}{7}$
3.	$\dfrac{1}{12}$	15.	$\dfrac{2}{7}$
4.	$\dfrac{1}{10}$	16.	$\dfrac{2}{9}$
5.	$\dfrac{1}{20}$	17.	$\dfrac{1}{6}$
6.	$\dfrac{1}{42}$	18.	$\dfrac{1}{12}$
7.	$\dfrac{1}{4}$	19.	$\dfrac{1}{4}$
8.	$\dfrac{1}{5}$	20.	$\dfrac{2}{25}$
9.	$\dfrac{1}{7}$	21.	$\dfrac{1}{9}$
10.	$\dfrac{1}{7}$	22.	$\dfrac{1}{24}$
11.	$\dfrac{2}{5}$	23.	$\dfrac{1}{36}$
12.	$\dfrac{1}{12}$	24.	$\dfrac{2}{27}$

Math Sprints 5

1.	$\dfrac{1}{4}$	13.	$\dfrac{1}{12}$
2.	$\dfrac{1}{6}$	14.	$\dfrac{2}{7}$
3.	$\dfrac{1}{12}$	15.	$\dfrac{2}{5}$
4.	$\dfrac{1}{6}$	16.	$\dfrac{2}{9}$
5.	$\dfrac{1}{20}$	17.	$\dfrac{1}{10}$
6.	$\dfrac{1}{35}$	18.	$\dfrac{1}{12}$
7.	$\dfrac{1}{7}$	19.	$\dfrac{3}{8}$
8.	$\dfrac{2}{5}$	20.	$\dfrac{3}{25}$
9.	$\dfrac{1}{9}$	21.	$\dfrac{1}{9}$
10.	$\dfrac{1}{5}$	22.	$\dfrac{1}{24}$
11.	$\dfrac{2}{5}$	23.	$\dfrac{1}{24}$
12.	$\dfrac{1}{18}$	24.	$\dfrac{2}{27}$

1.	2 : 4	16.	49 : 21
2.	2 : 8	17.	8 : 72
3.	3 : 9	18.	90 : 20
4.	20 : 4	19.	33 : 110
5.	4 : 6	20.	90 : 63
6.	36 : 3	21.	18 : 99
7.	9 : 12	22.	12 : 144
8.	20 : 25	23.	72 : 42
9.	8 : 20	24.	40 : 96
10.	9 : 15	25.	42 : 91
11.	28 : 8	26.	3 : 6 : 9
12.	21 : 56	27.	10 : 5 : 20
13.	18 : 42	28.	36 : 12 : 6
14.	56 : 35	29.	24 : 32 : 16
15.	25 : 20	30.	45 : 18 : 54

513 A   Simplify.   Second Half

1.	1 : 2	16.	8 : 72
2.	3 : 9	17.	49 : 21
3.	2 : 8	18.	18 : 4
4.	15 : 3	19.	30 : 100
5.	4 : 6	20.	60 : 42
6.	24 : 2	21.	18 : 99
7.	6 : 8	22.	8 : 36
8.	16 : 20	23.	72 : 6
9.	6 : 15	24.	56 : 96
10.	9 : 15	25.	24 : 52
11.	21 : 6	26.	2 : 4 : 6
12.	18 : 42	27.	10 : 5 : 25
13.	21 : 56	28.	36 : 24 : 6
14.	48 : 30	29.	32 : 16 : 24
15.	30 : 24	30.	18 : 54 : 45

| | 513 B | | Simplify. | | First Half |

1.	3 : 6	16.	56 : 24
2.	4 : 16	17.	9 : 81
3.	5 : 15	18.	99 : 22
4.	25 : 5	19.	36 : 120
5.	6 : 9	20.	90 : 63
6.	48 : 4	21.	24 : 132
7.	12 : 16	22.	13 : 156
8.	24 : 30	23.	96 : 56
9.	14 : 35	24.	45 : 108
10.	18 : 30	25.	48 : 104
11.	49 : 14	26.	4 : 8 : 12
12.	24 : 64	27.	12 : 6 : 24
13.	18 : 42	28.	42 : 14 : 7
14.	64 : 40	29.	27 : 36 : 18
15.	35 : 28	30.	60 : 24 : 72

513 B   Simplify.   Second Half

1.	3 : 6	16.	9 : 81
2.	5 : 15	17.	56 : 24
3.	2 : 8	18.	27 : 6
4.	20 : 4	19.	27 : 90
5.	6 : 9	20.	70 : 49
6.	36 : 3	21.	24 : 132
7.	12 : 16	22.	10 : 45
8.	28 : 35	23.	96 : 8
9.	14 : 35	24.	63 : 108
10.	15 : 25	25.	30 : 65
11.	42 : 12	26.	3 : 6 : 9
12.	18 : 42	27.	12 : 6 : 30
13.	24 : 64	28.	42 : 28 : 7
14.	56 : 35	29.	36 : 18 : 27
15.	40 : 32	30.	24 : 72 : 60

513 A & B		Answer Sheet		First Half

1.	1 : 2	16.	7 : 3
2.	1 : 4	17.	1 : 9
3.	1 : 3	18.	9 : 2
4.	5 : 1	19.	3 : 10
5.	2 : 3	20.	10 : 7
6.	12 : 1	21.	2 : 11
7.	3 : 4	22.	1 : 12
8.	4 : 5	23.	12 : 7
9.	2 : 5	24.	5 : 12
10.	3 : 5	25.	6 : 13
11.	7 : 2	26.	1 : 2 : 3
12.	3 : 8	27.	2 : 1 : 4
13.	3 : 7	28.	6 : 2 : 1
14.	8 : 5	29.	3 : 4 : 2
15.	5 : 4	30.	5 : 2 : 6

513 A & B   Answer Sheet   Second Half

1.	1 : 2	16.	1 : 9
2.	1 : 3	17.	7 : 3
3.	1 : 4	18.	9 : 2
4.	5 : 1	19.	3 : 10
5.	2 : 3	20.	10 : 7
6.	12 : 1	21.	2 : 11
7.	3 : 4	22.	2 : 9
8.	4 : 5	23.	12 : 1
9.	2 : 5	24.	7 : 12
10.	3 : 5	25.	6 : 13
11.	7 : 2	26.	1 : 2 : 3
12.	3 : 7	27.	2 : 1 : 5
13.	3 : 8	28.	6 : 4 : 1
14.	8 : 5	29.	4 : 2 : 3
15.	5 : 4	30.	2 : 6 : 5

	514 A		Simplify.		First Half

1.	2 : 4	16.	10 : 2
2.	2 : 6	17.	9 : 3
3.	2 : 10	18.	10 : 80
4.	6 : 3	19.	72 : 8
5.	12 : 4	20.	100 : 10
6.	8 : 12	21.	18 : 63
7.	9 : 6	22.	27 : 63
8.	20 : 15	23.	49 : 28
9.	5 : 15	24.	56 : 21
10.	6 : 30	25.	84 : 49
11.	12 : 30	26.	40 : 96
12.	30 : 40	27.	3 : 9 : 15
13.	30 : 18	28.	25 : 10 : 15
14.	15 : 12	29.	18 : 72 : 6
15.	3 : 18	30.	48 : 16 : 40

314 A   Simplify.   Second Half

1.	3 : 6	16.	2 : 10
2.	2 : 8	17.	70 : 10
3.	2 : 10	18.	24 : 3
4.	6 : 2	19.	72 : 8
5.	4 : 2	20.	90 : 9
6.	6 : 9	21.	18 : 63
7.	9 : 6	22.	18 : 48
8.	5 : 15	23.	49 : 28
9.	20 : 15	24.	49 : 21
10.	30 : 6	25.	40 : 96
11.	8 : 20	26.	84 : 49
12.	40 : 30	27.	2 : 6 : 10
13.	30 : 18	28.	10 : 20 : 15
14.	16 : 20	29.	12 : 72 : 6
15.	4 : 24	30.	40 : 16 : 48

514 B		Simplify.		First Half

1.	3 : 6	16.	40 : 8	
2.	3 : 9	17.	30 : 10	
3.	4 : 20	18.	9 : 72	
4.	8 : 4	19.	81 : 9	
5.	15 : 5	20.	110 : 11	
6.	10 : 15	21.	24 : 84	
7.	18 : 12	22.	36 : 84	
8.	24 : 18	23.	63 : 36	
9.	7 : 21	24.	64 : 24	
10.	8 : 40	25.	108 : 63	
11.	14 : 35	26.	40 : 96	
12.	21 : 28	27.	4 : 12 : 20	
13.	35 : 21	28.	30 : 12 : 18	
14.	40 : 32	29.	21 : 84 : 7	
15.	6 : 36	30.	54 : 18 : 45	

514 B   Simplify.   Second Half

1.	4 : 8	16.	8 : 40
2.	3 : 12	17.	21 : 3
3.	3 : 15	18.	32 : 4
4.	9 : 3	19.	81 : 9
5.	6 : 3	20.	110 : 11
6.	8 : 12	21.	24 : 84
7.	18 : 12	22.	27 : 72
8.	7 : 21	23.	56 : 32
9.	24 : 18	24.	56 : 24
10.	40 : 8	25.	40 : 96
11.	12 : 30	26.	108 : 63
12.	28 : 21	27.	3 : 9 : 15
13.	35 : 21	28.	12 : 24 : 18
14.	24 : 30	29.	14 : 84 : 7
15.	7 : 42	30.	45 : 18 : 54

Math Sprints 5

1.	1 : 2	16.	5 : 1
2.	1 : 3	17.	3 : 1
3.	1 : 5	18.	1 : 8
4.	2 : 1	19.	9 : 1
5.	3 : 1	20.	10 : 1
6.	2 : 3	21.	2 : 7
7.	3 : 2	22.	3 : 7
8.	4 : 3	23.	7 : 4
9.	1 : 3	24.	8 : 3
10.	1 : 5	25.	12 : 7
11.	2 : 5	26.	5 : 12
12.	3 : 4	27.	1 : 3 : 5
13.	5 : 3	28.	5 : 2 : 3
14.	5 : 4	29.	3 : 12 : 1
15.	1 : 6	30.	6 : 2 : 5

Math Sprints 5

1.	1 : 2	16.	1 : 5
2.	1 : 4	17.	7 : 1
3.	1 : 5	18.	8 : 1
4.	3 : 1	19.	9 : 1
5.	2 : 1	20.	10 : 1
6.	2 : 3	21.	2 : 7
7.	3 : 2	22.	3 : 8
8.	1 : 3	23.	7 : 4
9.	4 : 3	24.	7 : 3
10.	5 : 1	25.	5 : 12
11.	2 : 5	26.	12 : 7
12.	4 : 3	27.	1 : 3 : 5
13.	5 : 3	28.	2 : 4 : 3
14.	4 : 5	29.	2 : 12 : 1
15.	1 : 6	30.	5 : 2 : 6

515 A		Write the number as a decimal.		First Half

1.	3 tenths	16.	9 tenths 6 ones
2.	4 tenths	17.	7 ones 2 hundredths 1 tenth
3.	2 ones 1 tenth	18.	3 tenths 4 ones 5 tens
4.	4 ones 3 tenths	19.	7 ones 4 hundreds 6 tenths
5.	1 ten 2 ones 5 tenths	20.	2 hundreds 3 tenths 3 ones
6.	2 tens 4 ones 7 tenths	21.	2 hundredths 4 thousandths
7.	8 ones 4 tenths	22.	3 ones 3 thousands
8.	10 ones 3 tenths	23.	7 tens 4 tenths 6 thousandths
9.	3 hundredths	24.	4 tenths 1 hundredth 3 thousandths
10.	4 tenths 3 hundredths	25.	5 thousandths 3 ones
11.	7 ones 4 hundredths	26.	3 hundredths 5 tenths 2 tens
12.	3 tens 7 tenths	27.	3 thousandths 2 tens 4 tenths
13.	3 tens 7 hundredths	28.	5 hundredths 10 ones 3 tenths
14.	5 hundreds 2 hundredths	29.	10 hundredths
15.	30 ones 2 tenths	30.	40 thousandths

515 A   Write the number as a decimal.   Second Half

1.	2 tenths	16.	10 tenths
2.	5 tenths	17.	4 ones 6 hundredths 1 tenth
3.	3 ones 1 tenth	18.	9 tenths 1 one 3 tens
4.	4 ones 2 tenths	19.	4 ones 9 hundreds 6 tenths
5.	1 ten 1 one 3 tenths	20.	2 hundreds 3 tenths 3 ones
6.	2 tens 4 ones 6 tenths	21.	8 hundredths 2 thousandths
7.	8 ones 4 tenths	22.	6 ones 3 thousands
8.	10 ones 3 tenths	23.	9 tens 1 tenths 2 thousandths
9.	4 hundredths	24.	7 tenths 1 hundredth 2 thousandths
10.	2 tenths 8 hundredths	25.	7 thousandths 4 ones
11.	2 ones 3 hundredths	26.	3 hundredths 5 tenths 2 tens
12.	5 tens 7 tenths	27.	1 thousandths 9 tens 6 tenths
13.	3 tens 7 hundredths	28.	7 hundredths 20 ones 4 tenths
14.	5 hundreds	29.	10 hundredths
15.	5 hundreds 3 hundredths	30.	50 thousandths

Write the number as a decimal.

1.	3 tenths	16.	5 ones 19 tenths
2.	4 tenths	17.	7 ones 12 hundredths
3.	1 tenth 2 ones	18.	4 tens 14 ones 3 tenths
4.	3 tenths 4 ones	19.	5 ones 4 hundreds 26 tenths
5.	2 ones 5 tenths 1 ten	20.	2 hundreds 13 tenths 2 ones
6.	7 tenths 2 tens 4 ones	21.	4 thousandths 2 hundredths
7.	7 ones 14 tenths	22.	3 thousands 30 tenths
8.	3 tenths 10 ones	23.	7 tens 6 thousandths 4 tenths
9.	3 hundredths	24.	1 hundredth 3 thousandths 4 tenths
10.	3 hundredths 4 tenths	25.	5 thousandths 2 ones 10 tenths
11.	4 hundredths 7 ones	26.	4 tenths 13 hundredths 2 tens
12.	6 tenths 3 tens 10 hundredths	27.	3 thousandths 2 tens 40 hundredths
13.	7 hundredths 3 tens	28.	9 ones 13 tenths 5 hundredths
14.	2 hundredths 5 hundreds	29.	100 thousandths
15.	29 ones 12 tenths	30.	40 thousandths

515 B Write the number as a decimal. Second Half

1.	2 tenths	16.	10 tenths
2.	5 tenths	17.	4 ones 16 hundredths
3.	1 tenth 3 ones	18.	2 tens 11 ones 9 tenths
4.	2 tenths 4 ones	19.	3 ones 9 hundreds 16 tenths
5.	1 ones 3 tenths 1 ten	20.	2 hundreds 13 tenths 2 ones
6.	6 tenths 2 tens 4 ones	21.	2 thousandths 8 hundredths
7.	7 ones 14 tenths	22.	3 thousands 60 tenths
8.	3 tenths 1 ten	23.	9 tens 2 thousandths 1 tenth
9.	4 hundredths	24.	1 hundredth 2 thousandths 7 tenths
10.	8 hundredths 2 tenths	25.	7 thousandths 3 ones 10 tenths
11.	3 hundredths 2 ones	26.	4 tenths 13 hundredths 2 tens
12.	5 ones 70 hundredths	27.	1 thousandths 9 tens 60 hundredths
13.	7 hundredths 3 tens	28.	19 ones 14 tenths 7 hundredths
14.	50 tens	29.	100 thousandths
15.	3 hundredths 5 hundreds	30.	50 thousandths

1.	0.3	16.	6.9
2.	0.4	17.	7.12
3.	2.1	18.	54.3
4.	4.3	19.	407.6
5.	12.5	20.	203.3
6.	24.7	21.	0.024
7.	8.4	22.	3,003
8.	10.3	23.	70.406
9.	0.03	24.	0.413
10.	0.43	25.	3.005
11.	7.04	26.	20.53
12.	30.7	27.	20.403
13.	30.07	28.	10.35
14.	500.02	29.	0.1
15.	30.2	30.	0.04

1.	0.2	16.	1
2.	0.5	17.	4.16
3.	3.1	18.	31.9
4.	4.2	19.	904.6
5.	11.3	20.	203.3
6.	24.6	21.	0.082
7.	8.4	22.	3,006
8.	10.3	23.	90.102
9.	0.04	24.	0.712
10.	0.28	25.	4.007
11.	2.03	26.	20.53
12.	5.7	27.	90.601
13.	30.07	28.	20.47
14.	500	29.	0.1
15.	500.03	30.	0.05

516 A   Fill in the blank with >, <, or =.   First Half

1.	3.2 _____ 2.3	16.	1.896 _____ 1.986
2.	5.62 _____ 5.26	17.	14.231 _____ 14.132
3.	63.42 _____ 63.24	18.	3.42 _____ 3.042
4.	108.36 _____ 108.63	19.	5.389 _____ 5.942
5.	7.9 _____ 7.09	20.	10.081 _____ 10.801
6.	24.018 _____ 24.018	21.	192.36 _____ 192.63
7.	52.63 _____ 52.93	22.	8.459 _____ 8.954
8.	256.41 _____ 256.14	23.	3.281 _____ 3.28
9.	6.98 _____ 6.2	24.	4.9 _____ 4.89
10.	7.281 _____ 7.282	25.	2 + 0.1 + 0.02 _____ 0.02 + 0.1 + 2
11.	1.009 _____ 1.01	26.	0.6 + 0.05 _____ 0.5 + 0.06
12.	24.35 _____ 24.349	27.	0.21 + 0.005 _____ 0.25 + 0.001
13.	12.78 _____ 12.781	28.	1.23 + 0.11 _____ 1.22 + 0.12
14.	2.49 _____ 3.49	29.	7.41 + 0.011 _____ 6.5 + 1.021
15.	63.182 _____ 63.187	30.	12.02 + 0.6 _____ 11.01 + 1.59

Math Sprints 5

Fill in the blank with >, <, or =. Second Half

1.	9.1 _____ 1.9	16.	1.896 _____ 1.986
2.	5.56 _____ 5.52	17.	14.231 _____ 14.132
3.	63.42 _____ 63.24	18.	3.42 _____ 3.042
4.	201.36 _____ 201.63	19.	6.271 _____ 6.271 + 1
5.	8.9 _____ 8.09	20.	14.032 _____ 14.302
6.	32.018 _____ 32.018	21.	192.36 _____ 192.63
7.	11.23 _____ 11.32	22.	4.359 _____ 4.593
8.	130.4 _____ 130.1	23.	3.281 _____ 3.28
9.	6.98 _____ 6.2	24.	4.9 _____ 4.89
10.	4.281 _____ 4.282	25.	2 + 0.1 + 0.02 _____ 0.02 + 0.1 + 2
11.	2.009 _____ 2.01	26.	0.7 + 0.05 _____ 0.5 + 0.07
12.	16.35 _____ 16.349	27.	0.26 + 0.009 _____ 0.29 + 0.006
13.	12.49 _____ 12.491	28.	3.59 + 0.01 _____ 3.58 + 0.02
14.	2.5 _____ 3.49	29.	4.31 + 0.011 _____ 3.5 + 1.021
15.	39.182 _____ 39.187	30.	13.07 + 0.05 _____ 11.01 + 1.59

Math Sprints 5

Fill in the blank with >, <, or =.

1.	3.2 _____ 2.3	16.	1.896 _____ 1 + 0.9 + 0.08 + 0.006
2.	5.92 _____ 5.29	17.	41.414 _____ 41.141
3.	6.342 _____ 6.324	18.	3.42 _____ 3 + 0.042
4.	10.836 _____ 10.863	19.	5.492 _____ 5.942
5.	0.79 _____ 0.709	20.	10.081 _____ 10 + 0.801
6.	24.018 _____ 24 + 0.018	21.	190 + 2.36 _____ 100 + 92.63
7.	5.263 _____ 5.293	22.	8.4 + 0.059 _____ 8.95 + 0.004
8.	256.418 _____ 256.148	23.	3.2 + 0.081 _____ 3.2 + 0.08
9.	0.698 _____ 0.62	24.	4.4 + 0.5 _____ 4.89
10.	7.281 _____ 7.282	25.	2 + 0.06 + 0.06 _____ 2.07 + 0.05
11.	213.009 _____ 213.01	26.	0.32 + 0.33 _____ 0.23 + 0.33
12.	24 + 0.35 _____ 24 + 0.349	27.	0.21 + 0.005 _____ 0.25 + 0.001
13.	12 + 0.78 _____ 12 + 0.781	28.	0.59 + 0.62 _____ 0.49 + 0.72
14.	1.49 + 1.01 _____ 1.4 + 2.09	29.	215.06 + 21.5 _____ 121.5 + 131.06
15.	63 + 0.182 _____ 63 + 0.18 + 0.007	30.	1.842 + 0.999 _____ 2.181 + 0.034

Math Sprints 5

1.	9.1 _____ 1.9	16.	1.396 _____ 1 + 0.9 + 0.03 + 0.006
2.	5.95 _____ 5.59	17.	41.414 _____ 41.141
3.	6.342 _____ 6.324	18.	3.42 _____ 3 + 0.042
4.	20.147 _____ 20.174	19.	6.271 _____ 6.721
5.	0.89 _____ 0.809	20.	14.023 _____ 14 + 0.302
6.	32.018 _____ 32 + 0.018	21.	190 + 2.36 _____ 100 + 92.63
7.	1.123 _____ 1.132	22.	4.3 + 0.059 _____ 4.95 + 0.004
8.	130.41 _____ 130.14	23.	3.2 + 0.081 _____ 3.2 + 0.08
9.	0.698 _____ 0.62	24.	4.4 + 0.5 _____ 4.89
10.	4.281 _____ 4 + 0.282	25.	2 + 0.04 + 0.04 _____ 2.03 + 0.05
11.	171.009 _____ 171.01	26.	0.32 + 0.43 _____ 0.23 + 0.34
12.	56 + 0.35 _____ 56 + 0.349	27.	0.26 + 0.009 _____ 0.29 + 0.006
13.	12 + 0.49 _____ 12 + 0.491	28.	3.29 + 0.31 _____ 3.58 + 0.02
14.	1.5 + 1 _____ 3.4 + 0.09	29.	215.06 + 21.5 _____ 121.5 + 131.06
15.	39 + 0.182 _____ 39 + 0.18 + 0.007	30.	1.361 + 0.999 _____ 0.999 + 0.999

516 A & B   Answer Sheet   First Half

1.	>	16.	<
2.	>	17.	>
3.	>	18.	>
4.	<	19.	<
5.	>	20.	<
6.	=	21.	<
7.	<	22.	<
8.	>	23.	>
9.	>	24.	>
10.	<	25.	=
11.	<	26.	>
12.	>	27.	<
13.	<	28.	=
14.	<	29.	<
15.	<	30.	>

516 A & B   Answer Sheet   Second Half

1.	>	16.	<
2.	>	17.	>
3.	>	18.	>
4.	<	19.	<
5.	>	20.	<
6.	=	21.	<
7.	<	22.	<
8.	>	23.	>
9.	>	24.	>
10.	<	25.	=
11.	<	26.	>
12.	>	27.	<
13.	<	28.	=
14.	<	29.	<
15.	<	30.	>

517 A — Round to the nearest hundredth. — First Half

1.	1.021	14.	5.955
2.	7.143	15.	12.815
3.	10.059	16.	12.805
4.	3.614	17.	15.304
5.	9.219	18.	71.135
6.	14.964	19.	4.895
7.	21.438	20.	10.931
8.	6.744	21.	23.699
9.	54.745	22.	67.395
10.	29.893	23.	104.624
11.	32.181	24.	42.396
12.	63.952	25.	23.998
13.	8.038	26.	51.994

Round to the nearest hundredth.

1.	2.031	14.	5.955
2.	7.143	15.	15.315
3.	17.039	16.	19.305
4.	5.314	17.	27.304
5.	6.439	18.	71.135
6.	14.964	19.	5.896
7.	22.568	20.	30.931
8.	6.744	21.	56.699
9.	45.325	22.	67.395
10.	29.893	23.	251.314
11.	19.282	24.	96.496
12.	63.952	25.	23.998
13.	9.048	26.	99.995

517 B   Round the sum to the nearest hundredth.   First Half

1.	1.02 + 0.001	14.	5.944 + 0.011
2.	7.14 + 0.003	15.	12.809 + 0.006
3.	10.05 + 0.009	16.	12.301 + 0.504
4.	3.61 + 0.004	17.	15.299 + 0.003
5.	9.209 + 0.01	18.	71.11 + 0.025
6.	14.954 + 0.01	19.	4.87 + 0.025
7.	21.434 + 0.004	20.	10.811 + 0.12
8.	6.704 + 0.04	21.	23.666 + 0.033
9.	54.744 + 0.001	22.	67.320 + 0.075
10.	29.843 + 0.05	23.	104.612 + 0.012
11.	32.131 + 0.05	24.	40.363 + 2.033
12.	63.922 + 0.03	25.	20.934 + 3.064
13.	8.035 + 0.003	26.	26.994 + 25

517 B　　　　Round the sum to the nearest hundredth.　　　　Second Half

1.	2.03 + 0.001	14.	5.944 + 0.011
2.	7.14 + 0.003	15.	15.309 + 0.006
3.	17.03 + 0.009	16.	19.204 + 0.101
4.	5.31 + 0.004	17.	27.299 + 0.003
5.	6.409 + 0.03	18.	71.11 + 0.025
6.	14.954 + 0.01	19.	5.87 + 0.026
7.	22.564 + 0.004	20.	30.811 + 0.12
8.	6.704 + 0.04	21.	56.665 + 0.034
9.	45.324 + 0.001	22.	67.320 + 0.075
10.	29.843 + 0.05	23.	251.302 + 0.012
11.	19.28 + 0.002	24.	94.461 + 2.035
12.	63.922 + 0.03	25.	20.934 + 3.064
13.	9.045 + 0.003	26.	99.994 + 0.001

517 A & B		Answer Sheet		First Half
1.	1.02	14.	5.96	
2.	7.14	15.	12.82	
3.	10.06	16.	12.81	
4.	3.61	17.	15.30	
5.	9.22	18.	71.14	
6.	14.96	19.	4.9	
7.	21.44	20.	10.93	
8.	6.74	21.	23.7	
9.	54.75	22.	67.4	
10.	29.89	23.	104.62	
11.	32.18	24.	42.4	
12.	63.95	25.	24	
13.	8.04	26.	51.99	

517 A & B   Answer Sheet   Second Half

1.	2.03	14.	5.96
2.	7.14	15.	15.32
3.	17.04	16.	19.31
4.	5.31	17.	27.3
5.	6.44	18.	71.14
6.	14.96	19.	5.9
7.	22.57	20.	30.93
8.	6.74	21.	56.7
9.	45.33	22.	67.4
10.	29.89	23.	251.31
11.	19.28	24.	96.5
12.	63.95	25.	24
13.	9.05	26.	100

518 A		Add.		First Half

1.	2.3 + 0.1 =	13.	5.54 + 3.45 =
2.	4.5 + 0.4 =	14.	3.36 + 6.21 =
3.	9.5 + 0.6 =	15.	8.74 + 3.42 =
4.	3.7 + 4.2 =	16.	4.361 + 0.03 =
5.	8.3 + 2.9 =	17.	1.246 + 0.13 =
6.	3.64 + 0.3 =	18.	0.015 + 0.23 =
7.	5.37 + 0.9 =	19.	4.076 + 0.03 =
8.	0.98 + 0.99 =	20.	10.342 + 0.19 =
9.	3.9 + 5.23 =	21.	3.721 + 1.09 =
10.	35.4 + 7.35 =	22.	6.921 + 2.11 =
11.	0.72 + 0.62 =	23.	1.439 + 10.27 =
12.	0.09 + 0.21 =	24.	0.999 + 0.421 =

518 A		Add.		Second Half

1.	2.3 + 0.2 =		13.	5.54 + 4.45 =
2.	5.5 + 0.4 =		14.	2.36 + 6.21 =
3.	4.5 + 0.6 =		15.	9.74 + 3.43 =
4.	2.7 + 4.2 =		16.	4.361 + 0.03 =
5.	4.2 + 2.9 =		17.	2.246 + 0.13 =
6.	3.64 + 0.3 =		18.	0.075 + 0.13 =
7.	4.47 + 0.9 =		19.	9.072 + 0.03 =
8.	0.98 + 0.99 =		20.	7.342 + 0.19 =
9.	3.9 + 5.9 =		21.	1.721 + 1.09 =
10.	0.13 + 0.17 =		22.	6.921 + 3.11 =
11.	0.72 + 0.62 =		23.	0.439 + 10.27 =
12.	0.09 + 0.21 =		24.	0.999 + 0.421 =

	518 B		Subtract.		First Half

1.	$2.5 - 0.1 =$		13.	$9.43 - 0.44 =$
2.	$5.0 - 0.1 =$		14.	$11.56 - 1.99 =$
3.	$11.3 - 1.2 =$		15.	$16.14 - 3.98 =$
4.	$9.3 - 1.4 =$		16.	$4.441 - 0.05 =$
5.	$15.1 - 3.9 =$		17.	$1.796 - 0.42 =$
6.	$4.14 - 0.2 =$		18.	$1 - 0.755 =$
7.	$7.17 - 0.9 =$		19.	$4.946 - 0.84 =$
8.	$3.36 - 1.39 =$		20.	$10.612 - 0.08 =$
9.	$15.03 - 5.9 =$		21.	$5.801 - 0.99 =$
10.	$55.65 - 12.9 =$		22.	$12.021 - 2.99 =$
11.	$1.72 - 0.38 =$		23.	$15.259 - 3.55 =$
12.	$3.02 - 2.72 =$		24.	$1.437 - 0.017 =$

518 B		Subtract.		Second Half

1.	$2.6 - 0.1 =$		13.	$10.43 - 0.44 =$
2.	$6.0 - 0.1 =$		14.	$10.56 - 1.99 =$
3.	$6.3 - 1.2 =$		15.	$17.15 - 3.98 =$
4.	$8.3 - 1.4 =$		16.	$4.441 - 0.05 =$
5.	$11 - 3.9 =$		17.	$2.796 - 0.42 =$
6.	$4.14 - 0.2 =$		18.	$1 - 0.795 =$
7.	$6.27 - 0.9 =$		19.	$9.942 - 0.84 =$
8.	$3.36 - 1.39 =$		20.	$7.612 - 0.08 =$
9.	$15.7 - 5.9 =$		21.	$3.801 - 0.99 =$
10.	$40.1 - 39.8 =$		22.	$12.021 - 1.99 =$
11.	$1.72 - 0.38 =$		23.	$14.259 - 3.55 =$
12.	$3.02 - 2.72 =$		24.	$1.537 - 0.117 =$

1.	2.4	13.	8.99
2.	4.9	14.	9.57
3.	10.1	15.	12.16
4.	7.9	16.	4.391
5.	11.2	17.	1.376
6.	3.94	18.	0.245
7.	6.27	19.	4.106
8.	1.97	20.	10.532
9.	9.13	21.	4.811
10.	42.75	22.	9.031
11.	1.34	23.	11.709
12.	0.3	24.	1.42

1.	2.5	13.	9.99
2.	5.9	14.	8.57
3.	5.1	15.	13.17
4.	6.9	16.	4.391
5.	7.1	17.	2.376
6.	3.94	18.	0.205
7.	5.37	19.	9.102
8.	1.97	20.	7.532
9.	9.8	21.	2.811
10.	0.3	22.	10.031
11.	1.34	23.	10.709
12.	0.3	24.	1.42

519 A		Subtract.	First Half

1.	$1 - 0.4 =$	15.	$8.6 - 5.2 =$
2.	$2 - 0.4 =$	16.	$4.1 - 0.2 =$
3.	$3 - 0.4 =$	17.	$4.1 - 2 =$
4.	$3 - 0.7 =$	18.	$4.1 - 1.9 =$
5.	$3 - 1.7 =$	19.	$6.2 - 1 =$
6.	$3.1 - 0.9 =$	20.	$6.2 - 0.9 =$
7.	$4.1 - 0.8 =$	21.	$6.3 - 1.3 =$
8.	$5 - 0.6 =$	22.	$6.3 - 2.3 =$
9.	$5.2 - 0.6 =$	23.	$6.3 - 4.3 =$
10.	$6.2 - 0.6 =$	24.	$6.3 - 4.5 =$
11.	$6.2 - 1.6 =$	25.	$7.5 - 2.9 =$
12.	$9.5 - 0.9 =$	26.	$8 - 2.9 =$
13.	$9.7 - 1.9 =$	27.	$8.2 - 4.9 =$
14.	$9.7 - 2.9 =$	28.	$8.2 - 4.8 =$

519 A		Subtract.			Second Half
1.	$1 - 0.6 =$		15.	$7.6 - 5.2 =$	
2.	$2 - 0.6 =$		16.	$3.1 - 0.2 =$	
3.	$3 - 0.6 =$		17.	$3.1 - 2 =$	
4.	$3 - 0.7 =$		18.	$4.1 - 2.9 =$	
5.	$3 - 1.7 =$		19.	$6.2 - 4 =$	
6.	$4.1 - 0.9 =$		20.	$6.2 - 3.9 =$	
7.	$4.1 - 0.8 =$		21.	$6.3 - 1.3 =$	
8.	$5 - 2.1 =$		22.	$6.3 - 1.4 =$	
9.	$6.2 - 0.6 =$		23.	$7.3 - 4.3 =$	
10.	$7.2 - 0.6 =$		24.	$7.3 - 4.5 =$	
11.	$10.7 - 6.1 =$		25.	$8.5 - 2.9 =$	
12.	$10.7 - 2.1 =$		26.	$10 - 4.9 =$	
13.	$7.7 - 1.9 =$		27.	$18.2 - 14.9 =$	
14.	$7.7 - 2.9 =$		28.	$18.2 - 14.8 =$	

519 B		Subtract.		First Half

1.	$1 - 0.4 =$		15.	$8.6 - 5.2 =$
2.	$2 - 0.4 =$		16.	$8.1 - 4.2 =$
3.	$4 - 1.4 =$		17.	$8.1 - 6 =$
4.	$4 - 1.7 =$		18.	$8.1 - 5.9 =$
5.	$4 - 2.7 =$		19.	$12.1 - 6.9 =$
6.	$5.1 - 2.9 =$		20.	$12.1 - 6.8 =$
7.	$6.1 - 2.8 =$		21.	$9.1 - 4.1 =$
8.	$6 - 1.6 =$		22.	$9.1 - 5.1 =$
9.	$7.2 - 2.6 =$		23.	$9.1 - 7.1 =$
10.	$8.2 - 2.6 =$		24.	$9.1 - 7.3 =$
11.	$9.7 - 5.1 =$		25.	$11.5 - 6.9 =$
12.	$10.7 - 2.1 =$		26.	$10 - 4.9 =$
13.	$9.2 - 1.4 =$		27.	$14.2 - 10.9 =$
14.	$9.2 - 2.4 =$		28.	$6.1 - 2.7 =$

Math Sprints 5

1.	$1 - 0.6 =$		15.	$7.6 - 5.2 =$
2.	$2 - 0.6 =$		16.	$7.1 - 4.2 =$
3.	$4 - 1.6 =$		17.	$7.1 - 6 =$
4.	$4 - 1.7 =$		18.	$8.1 - 6.9 =$
5.	$4 - 2.7 =$		19.	$12.1 - 9.9 =$
6.	$6.1 - 2.9 =$		20.	$12.1 - 9.8 =$
7.	$6.1 - 2.8 =$		21.	$9.1 - 4.1 =$
8.	$6 - 3.1 =$		22.	$9.1 - 4.2 =$
9.	$6.9 - 1.3 =$		23.	$10.1 - 7.1 =$
10.	$7.9 - 1.3 =$		24.	$10.1 - 7.3 =$
11.	$8.2 - 3.6 =$		25.	$12.5 - 6.9 =$
12.	$9.5 - 0.9 =$		26.	$11 - 5.9 =$
13.	$7.2 - 1.4 =$		27.	$24.2 - 20.9 =$
14.	$7.2 - 2.4 =$		28.	$16.1 - 12.7 =$

519 A & B		Answer Sheet		First Half

1.	0.6	15.	3.4
2.	1.6	16.	3.9
3.	2.6	17.	2.1
4.	2.3	18.	2.2
5.	1.3	19.	5.2
6.	2.2	20.	5.3
7.	3.3	21.	5
8.	4.4	22.	4
9.	4.6	23.	2
10.	5.6	24.	1.8
11.	4.6	25.	4.6
12.	8.6	26.	5.1
13.	7.8	27.	3.3
14.	6.8	28.	3.4

519 A & B   Answer Sheet   Second Half

1.	0.4	15.	2.4
2.	1.4	16.	2.9
3.	2.4	17.	1.1
4.	2.3	18.	1.2
5.	1.3	19.	2.2
6.	3.2	20.	2.3
7.	3.3	21.	5
8.	2.9	22.	4.9
9.	5.6	23.	3
10.	6.6	24.	2.8
11.	4.6	25.	5.6
12.	8.6	26.	5.1
13.	5.8	27.	3.3
14.	4.8	28.	3.4

520 A		Subtract.	First Half

1.	8 – 4 =	12.	23 – 0.9 =
2.	1.8 – 0.4 =	13.	12 – 8 =
3.	7.8 – 0.4 =	14.	1.2 – 0.8 =
4.	5.78 – 0.04 =	15.	79 – 35 =
5.	57.8 – 4 =	16.	0.79 – 0.35 =
6.	3.8 – 0.2 =	17.	47 – 38 =
7.	3.8 – 1.2 =	18.	1.47 – 1.38 =
8.	3.38 – 0.02 =	19.	90 – 59 =
9.	7 – 1 =	20.	9 – 5.9 =
10.	7 – 0.9 =	21.	2.2 – 1.5 =
11.	13 – 0.9 =	22.	1.72 – 0.93 =

Math Sprints 5

520 A		Subtract.		Second Half

1.	6 − 2 =		12.	13 − 0.9 =
2.	0.6 − 0.2 =		13.	14 − 8 =
3.	7.6 − 0.2 =		14.	1.4 − 0.8 =
4.	2.76 − 0.02 =		15.	79 − 36 =
5.	57.8 − 3 =		16.	0.79 − 0.36 =
6.	9.8 − 0.2 =		17.	57 − 38 =
7.	9.8 − 1.2 =		18.	1.57 − 1.38 =
8.	9.38 − 0.02 =		19.	100 − 59 =
9.	8 − 1 =		20.	10 − 5.9 =
10.	8 − 0.9 =		21.	2.2 − 1.5 =
11.	19 − 0.9 =		22.	1.72 − 0.93 =

520 B	Subtract.	First Half

1.	18 – 14 =	12.	33 – 10.9 =
2.	1.8 – 0.4 =	13.	22 – 18 =
3.	8.8 – 1.4 =	14.	2.2 – 1.8 =
4.	6.78 – 1.04 =	15.	179 – 135 =
5.	67.8 – 14 =	16.	1.79 – 1.35 =
6.	4.8 – 1.2 =	17.	147 – 138 =
7.	4.8 – 2.2 =	18.	3.47 – 3.38 =
8.	3.48 – 0.12 =	19.	190 – 159 =
9.	7 – 1 =	20.	19 – 15.9 =
10.	7 – 0.9 =	21.	4.2 – 3.5 =
11.	14 – 1.9 =	22.	2.72 – 1.93 =

Math Sprints 5

1.	$16 - 12 =$	12.	$23 - 10.9 =$
2.	$1.6 - 1.2 =$	13.	$24 - 18 =$
3.	$8.6 - 1.2 =$	14.	$2.4 - 1.8 =$
4.	$3.76 - 1.02 =$	15.	$179 - 136 =$
5.	$67.8 - 13 =$	16.	$1.79 - 1.36 =$
6.	$10.8 - 1.2 =$	17.	$157 - 138 =$
7.	$10.8 - 2.2 =$	18.	$3.57 - 3.38 =$
8.	$9.48 - 0.12 =$	19.	$200 - 159 =$
9.	$8 - 1 =$	20.	$20 - 15.9 =$
10.	$8 - 0.9 =$	21.	$4.2 - 3.5 =$
11.	$20 - 1.9 =$	22.	$2.72 - 1.93 =$

1.	4	12.	22.1
2.	1.4	13.	4
3.	7.4	14.	0.4
4.	5.74	15.	44
5.	53.8	16.	0.44
6.	3.6	17.	9
7.	2.6	18.	0.09
8.	3.36	19.	31
9.	6	20.	3.1
10.	6.1	21.	0.7
11.	12.1	22.	0.79

Math Sprints 5

1.	4	12.	12.1
2.	0.4	13.	6
3.	7.4	14.	0.6
4.	2.74	15.	43
5.	54.8	16.	0.43
6.	9.6	17.	19
7.	8.6	18.	0.19
8.	9.36	19.	41
9.	7	20.	4.1
10.	7.1	21.	0.7
11.	18.1	22.	0.79

521 A		Fill in the blank.	**First Half**

1.	$1 + \underline{\hspace{2cm}} = 1.1$	13.	$2.51 + \underline{\hspace{2cm}} = 8.81$
2.	$1.1 + \underline{\hspace{2cm}} = 3.4$	14.	$3.6 + \underline{\hspace{2cm}} = 11.89$
3.	$0.3 + \underline{\hspace{2cm}} = 7$	15.	$0.17 + \underline{\hspace{2cm}} = 6.97$
4.	$5 + \underline{\hspace{2cm}} = 9.8$	16.	$0.12 + \underline{\hspace{2cm}} = 0.46$
5.	$0.4 + \underline{\hspace{2cm}} = 6.9$	17.	$0.43 + \underline{\hspace{2cm}} = 10$
6.	$0.1 + \underline{\hspace{2cm}} = 4$	18.	$1.21 + \underline{\hspace{2cm}} = 5.89$
7.	$3.7 + \underline{\hspace{2cm}} = 13.9$	19.	$3.15 + \underline{\hspace{2cm}} = 24.47$
8.	$2.3 + \underline{\hspace{2cm}} = 2.39$	20.	$1.99 + \underline{\hspace{2cm}} = 44.74$
9.	$0.03 + \underline{\hspace{2cm}} = 8.66$	21.	$9.24 + \underline{\hspace{2cm}} = 10.23$
10.	$0.1 + \underline{\hspace{2cm}} = 9.124$	22.	$8.25 + \underline{\hspace{2cm}} = 11.86$
11.	$0.002 + \underline{\hspace{2cm}} = 9.026$	23.	$4.6 + \underline{\hspace{2cm}} = 11.831$
12.	$3.015 + \underline{\hspace{2cm}} = 3.021$	24.	$1.9 + \underline{\hspace{2cm}} = 10.552$

521 A   Fill in the blank.   Second Half

1.	$1 + \underline{\hspace{2cm}} = 1.2$	13.	$3.51 + \underline{\hspace{2cm}} = 8.81$
2.	$1.1 + \underline{\hspace{2cm}} = 3.3$	14.	$3.6 + \underline{\hspace{2cm}} = 10.89$
3.	$0.3 + \underline{\hspace{2cm}} = 7$	15.	$0.17 + \underline{\hspace{2cm}} = 6.87$
4.	$4 + \underline{\hspace{2cm}} = 9.8$	16.	$0.12 + \underline{\hspace{2cm}} = 0.57$
5.	$0.5 + \underline{\hspace{2cm}} = 6.9$	17.	$0.43 + \underline{\hspace{2cm}} = 10$
6.	$0.1 + \underline{\hspace{2cm}} = 5$	18.	$2.21 + \underline{\hspace{2cm}} = 6.89$
7.	$4.7 + \underline{\hspace{2cm}} = 13.9$	19.	$3.15 + \underline{\hspace{2cm}} = 14.47$
8.	$3.3 + \underline{\hspace{2cm}} = 3.39$	20.	$1.99 + \underline{\hspace{2cm}} = 33.74$
9.	$0.03 + \underline{\hspace{2cm}} = 8.66$	21.	$5.67 + \underline{\hspace{2cm}} = 6.66$
10.	$0.1 + \underline{\hspace{2cm}} = 7.196$	22.	$10.35 + \underline{\hspace{2cm}} = 13.96$
11.	$0.003 + \underline{\hspace{2cm}} = 9.027$	23.	$3.7 + \underline{\hspace{2cm}} = 11.831$
12.	$3.015 + \underline{\hspace{2cm}} = 3.021$	24.	$1.9 + \underline{\hspace{2cm}} = 11.641$

Fill in the blank.

1.	$2 + \rule{3cm}{0.4pt} = 2.1$	13.	$4.71 + \rule{3cm}{0.4pt} = 11.01$
2.	$3.1 + \rule{3cm}{0.4pt} = 5.4$	14.	$2.9 + \rule{3cm}{0.4pt} = 11.19$
3.	$1.3 + \rule{3cm}{0.4pt} = 8$	15.	$3.92 + \rule{3cm}{0.4pt} = 10.72$
4.	$3.2 + \rule{3cm}{0.4pt} = 8$	16.	$0.48 + \rule{3cm}{0.4pt} = 0.82$
5.	$6.2 + \rule{3cm}{0.4pt} = 12.7$	17.	$0.43 + \rule{3cm}{0.4pt} = 10$
6.	$0.3 + \rule{3cm}{0.4pt} = 4.2$	18.	$1.22 + \rule{3cm}{0.4pt} = 5.9$
7.	$2.9 + \rule{3cm}{0.4pt} = 13.1$	19.	$3.19 + \rule{3cm}{0.4pt} = 24.51$
8.	$21.9 + \rule{3cm}{0.4pt} = 21.99$	20.	$1.98 + \rule{3cm}{0.4pt} = 44.73$
9.	$0.09 + \rule{3cm}{0.4pt} = 8.72$	21.	$14.56 + \rule{3cm}{0.4pt} = 15.55$
10.	$0.9 + \rule{3cm}{0.4pt} = 9.924$	22.	$4.09 + \rule{3cm}{0.4pt} = 7.7$
11.	$0.007 + \rule{3cm}{0.4pt} = 9.031$	23.	$3.24 + \rule{3cm}{0.4pt} = 10.471$
12.	$3.315 + \rule{3cm}{0.4pt} = 3.321$	24.	$2.9 + \rule{3cm}{0.4pt} = 11.552$

1.	2 + _____ = 2.2	13.	5.71 + _____ = 11.01
2.	3.1 + _____ = 5.3	14.	2.9 + _____ = 10.19
3.	1.3 + _____ = 8	15.	3.92 + _____ = 10.62
4.	2.2 + _____ = 8	16.	0.37 + _____ = 0.82
5.	6.3 + _____ = 12.7	17.	0.43 + _____ = 10
6.	0.3 + _____ = 5.2	18.	4.22 + _____ = 8.9
7.	3.9 + _____ = 13.1	19.	3.19 + _____ = 14.51
8.	31.9 + _____ = 31.99	20.	1.98 + _____ = 33.73
9.	0.09 + _____ = 8.72	21.	17.46 + _____ = 18.45
10.	0.9 + _____ = 7.996	22.	6.09 + _____ = 9.7
11.	0.006 + _____ = 9.03	23.	2.34 + _____ = 10.471
12.	3.315 + _____ = 3.321	24.	2.9 + _____ = 12.641

521 A & B		Answer Sheet		First Half

1.	0.1	13.	6.3
2.	2.3	14.	8.29
3.	6.7	15.	6.8
4.	4.8	16.	0.34
5.	6.5	17.	9.57
6.	3.9	18.	4.68
7.	10.2	19.	21.32
8.	0.09	20.	42.75
9.	8.63	21.	0.99
10.	9.024	22.	3.61
11.	9.024	23.	7.231
12.	0.006	24.	8.652

Math Sprints 5

1.	0.2	13.	5.3
2.	2.2	14.	7.29
3.	6.7	15.	6.7
4.	5.8	16.	0.45
5.	6.4	17.	9.57
6.	4.9	18.	4.68
7.	9.2	19.	11.32
8.	0.09	20.	31.75
9.	8.63	21.	0.99
10.	7.096	22.	3.61
11.	9.024	23.	8.131
12.	0.006	24.	9.741

1.	1 m = _____ cm	13.	5.7 m = _____ cm
2.	1 km = _____ m	14.	9.1 m = _____ cm
3.	_____ yd = 3 ft	15.	1.1 yd = _____ ft
4.	1 lb = _____ oz	16.	1.1 gal = _____ qt
5.	_____ qt = 4 c	17.	_____ gal = 4.8 qt
6.	2 yd = _____ ft	18.	0.75 m = _____ cm
7.	_____ lb = 32 oz	19.	_____ m = 365 cm
8.	1 gal = _____ qt	20.	0.45 kg = _____ g
9.	3 m = _____ cm	21.	3.23 kg = _____ g
10.	5 m 60 cm = _____ cm	22.	1.5 yd = _____ ft
11.	2 kg = _____ g	23.	1.5 ft = _____ in.
12.	1.4 m = _____ cm	24.	1.2 lb = _____ oz

Math Sprints 5

522 A   Fill in the blank with a decimal.   Second Half

1.	2 m = _____ cm	13.	6.8 m = _____ cm
2.	2 km = _____ m	14.	5.1 m = _____ cm
3.	_____ yd = 3 ft	15.	1.1 yd = _____ ft
4.	1 gal = _____ qt	16.	1.2 gal = _____ qt
5.	1 qt = _____ c	17.	_____ gal = 4.4 qt
6.	2 yd = _____ ft	18.	0.7 m = _____ cm
7.	_____ lb = 16 oz	19.	_____ m = 355 cm
8.	0.25 lb = _____ oz	20.	0.65 kg = _____ g
9.	4 m = _____ cm	21.	3.23 kg = _____ g
10.	7 m 40 cm = _____ cm	22.	2.3 yd = _____ ft
11.	3 kg = _____ g	23.	1.5 yd = _____ in.
12.	2.1 m = _____ cm	24.	1.3 lb = _____ oz

522 B   Fill in the blank with a decimal.   **First Half**

1.	1 m = _____ cm	13.	3.4 m + 2.3 m = _____ cm
2.	10 m = _____ cm	14.	8.9 m + 0.2 m = _____ cm
3.	_____ yd = 36 in.	15.	1.1 yd = _____ ft
4.	4 gal = _____ qt	16.	0.5 gal + 0.6 gal = _____ qt
5.	_____ qt = 4 c	17.	_____ qt = 4.8 c
6.	_____ yd = 18 ft	18.	1 m − 0.25 m = _____ cm
7.	_____ lb = 32 oz	19.	_____ m = 400 cm − 35 cm
8.	_____ qt = 8 pt	20.	1 kg − 0.55 kg = _____ g
9.	4 m = 100 cm + _____ cm	21.	4 kg − 0.77 kg = _____ g
10.	3 m + 2 m 60 cm = _____ cm	22.	2 yd − 0.5 yd = _____ ft
11.	1 kg 1000 g = _____ g	23.	2 ft − 0.5 ft = _____ in.
12.	1.4 m = _____ cm	24.	2 lb − 0.8 lb = _____ oz

Fill in the blank with a decimal.

1.	2 m = _____ cm	13.	3.2 m + 3.6 m = _____ cm
2.	20 m = _____ cm	14.	4.9 m + 0.2 m = _____ cm
3.	_____ yd = 36 in.	15.	1.1 yd = _____ ft
4.	_____ gal = 16 qt	16.	0.5 gal + 0.7 gal = _____ qt
5.	1 qt = _____ c	17.	_____ qt = 4.4 c
6.	_____ qt = 24 c	18.	1 m − 0.3 m = _____ cm
7.	_____ lb = 16 oz	19.	_____ m = 400 cm − 45 cm
8.	_____ qt = 8 pt	20.	1 kg − 0.35 kg = _____ g
9.	5 m = 100 cm + _____ cm	21.	4 kg − 0.77 kg = _____ g
10.	5 m + 2 m 40 cm = _____ cm	22.	3 yd − 0.7 yd = _____ ft
11.	2 kg 1000 g = _____ g	23.	2 yd − 0.5 yd = _____ in.
12.	2.1 m = _____ cm	24.	2 lb − 0.7 lb = _____ oz

522 A & B   Answer Sheet   First Half

1.	100	13.	570
2.	1000	14.	910
3.	1	15.	3.3
4.	16	16.	4.4
5.	1	17.	1.2
6.	6	18.	75
7.	2	19.	3.65
8.	4	20.	450
9.	300	21.	3.230
10.	560	22.	4.5
11.	2000	23.	18
12.	140	24.	19.2

522 A & B		Answer Sheet		Second Half

1.	200	13.	680	
2.	2000	14.	510	
3.	1	15.	3.3	
4.	4	16.	4.8	
5.	4	17.	1.1	
6.	6	18.	70	
7.	1	19.	3.55	
8.	4	20.	650	
9.	400	21.	3230	
10.	740	22.	6.9	
11.	3000	23.	54	
12.	210	24.	20.8	

523 A		Express as a decimal.			First Half

1.	$\dfrac{5}{10}$		15.	$\dfrac{23}{100}$	
2.	$\dfrac{1}{2}$		16.	$\dfrac{1}{25}$	
3.	$\dfrac{6}{10}$		17.	$\dfrac{3}{25}$	
4.	$\dfrac{7}{10}$		18.	$\dfrac{3}{50}$	
5.	$\dfrac{1}{10}$		19.	$\dfrac{43}{50}$	
6.	$\dfrac{75}{100}$		20.	$\dfrac{10}{25}$	
7.	$\dfrac{3}{4}$		21.	$\dfrac{20}{25}$	
8.	$\dfrac{1}{4}$		22.	$\dfrac{12}{25}$	
9.	$\dfrac{2}{10}$		23.	$1\dfrac{1}{10}$	
10.	$\dfrac{3}{5}$		24.	$2\dfrac{3}{10}$	
11.	$\dfrac{1}{2}$		25.	$7\dfrac{23}{100}$	
12.	$\dfrac{2}{5}$		26.	$3\dfrac{1}{4}$	
13.	$\dfrac{3}{10}$		27.	$4\dfrac{1}{2}$	
14.	$\dfrac{7}{10}$		28.	$10\dfrac{1}{5}$	

523 A		Express as a decimal.	Second Half

1.	$\dfrac{5}{10}$	15.	$\dfrac{21}{100}$
2.	$\dfrac{1}{2}$	16.	$\dfrac{1}{25}$
3.	$\dfrac{3}{10}$	17.	$\dfrac{2}{25}$
4.	$\dfrac{8}{10}$	18.	$\dfrac{4}{50}$
5.	$\dfrac{1}{10}$	19.	$\dfrac{23}{50}$
6.	$\dfrac{25}{100}$	20.	$\dfrac{20}{25}$
7.	$\dfrac{1}{4}$	21.	$\dfrac{11}{25}$
8.	$\dfrac{2}{4}$	22.	$\dfrac{12}{25}$
9.	$\dfrac{3}{10}$	23.	$1\dfrac{2}{10}$
10.	$\dfrac{1}{5}$	24.	$2\dfrac{3}{10}$
11.	$\dfrac{1}{2}$	25.	$8\dfrac{19}{100}$
12.	$\dfrac{2}{5}$	26.	$9\dfrac{3}{4}$
13.	$\dfrac{4}{10}$	27.	$2\dfrac{1}{2}$
14.	$\dfrac{9}{10}$	28.	$10\dfrac{1}{5}$

523 B   Express as a decimal.   First Half

1.	$\dfrac{5}{10}$		15.	$\dfrac{19}{100} + \dfrac{4}{100}$
2.	$\dfrac{1}{2}$		16.	$\dfrac{1}{25}$
3.	$\dfrac{4}{10} + \dfrac{2}{10}$		17.	$\dfrac{1}{25} + \dfrac{2}{25}$
4.	$\dfrac{3}{10} + \dfrac{4}{10}$		18.	$\dfrac{3}{50}$
5.	$\dfrac{1}{20} + \dfrac{1}{20}$		19.	$\dfrac{40}{50} + \dfrac{3}{50}$
6.	$\dfrac{50}{100} + \dfrac{25}{100}$		20.	$\dfrac{6}{25} + \dfrac{4}{25}$
7.	$\dfrac{1}{2} + \dfrac{1}{4}$		21.	$\dfrac{12}{25} + \dfrac{8}{25}$
8.	$\dfrac{1}{8} + \dfrac{1}{8}$		22.	$\dfrac{5}{25} + \dfrac{7}{25}$
9.	$\dfrac{4}{20}$		23.	$\dfrac{11}{10}$
10.	$\dfrac{2}{5} + \dfrac{1}{5}$		24.	$\dfrac{23}{10}$
11.	$\dfrac{1}{4} + \dfrac{1}{4}$		25.	$\dfrac{723}{100}$
12.	$\dfrac{1}{5} + \dfrac{1}{5}$		26.	$3\dfrac{2}{8}$
13.	$\dfrac{2}{10} + \dfrac{1}{10}$		27.	$4\dfrac{2}{4}$
14.	$\dfrac{5}{10} + \dfrac{2}{10}$		28.	$\dfrac{51}{5}$

Express as a decimal.

1.	$\dfrac{5}{10}$		15.	$\dfrac{19}{100} + \dfrac{2}{100}$
2.	$\dfrac{1}{2}$		16.	$\dfrac{1}{25}$
3.	$\dfrac{1}{10} + \dfrac{2}{10}$		17.	$\dfrac{1}{25} + \dfrac{1}{25}$
4.	$\dfrac{3}{10} + \dfrac{5}{10}$		18.	$\dfrac{4}{50}$
5.	$\dfrac{1}{20} + \dfrac{1}{20}$		19.	$\dfrac{20}{50} + \dfrac{3}{50}$
6.	$\dfrac{25}{100}$		20.	$\dfrac{16}{25} + \dfrac{4}{25}$
7.	$\dfrac{1}{8} + \dfrac{1}{8}$		21.	$\dfrac{11}{25}$
8.	$\dfrac{1}{4} + \dfrac{1}{4}$		22.	$\dfrac{5}{25} + \dfrac{7}{25}$
9.	$\dfrac{6}{20}$		23.	$\dfrac{12}{10}$
10.	$\dfrac{1}{5}$		24.	$\dfrac{23}{10}$
11.	$\dfrac{1}{4} + \dfrac{1}{4}$		25.	$\dfrac{819}{100}$
12.	$\dfrac{1}{5} + \dfrac{1}{5}$		26.	$9\dfrac{6}{8}$
13.	$\dfrac{3}{10} + \dfrac{1}{10}$		27.	$2\dfrac{3}{6}$
14.	$\dfrac{5}{10} + \dfrac{4}{10}$		28.	$\dfrac{51}{5}$

523 A & B		Answer Sheet		First Half
1.	0.5	15.		0.23
2.	0.5	16.		0.04
3.	0.6	17.		0.12
4.	0.7	18.		0.06
5.	0.1	19.		0.86
6.	0.75	20.		0.4
7.	0.75	21.		0.8
8.	0.25	22.		0.48
9.	0.2	23.		1.1
10.	0.6	24.		2.3
11.	0.5	25.		7.23
12.	0.4	26.		3.25
13.	0.3	27.		4.5
14.	0.7	28.		10.2

1.	0.5	15.	0.21
2.	0.5	16.	0.04
3.	0.3	17.	0.08
4.	0.8	18.	0.08
5.	0.1	19.	0.46
6.	0.25	20.	0.8
7.	0.25	21.	0.44
8.	0.5	22.	0.48
9.	0.3	23.	1.2
10.	0.2	24.	2.3
11.	0.5	25.	8.19
12.	0.4	26.	9.75
13.	0.4	27.	2.5
14.	0.9	28.	10.2

524 A Express as a decimal. First Half

1.	$\dfrac{1}{10}$		15.	$\dfrac{2}{5}$
2.	$\dfrac{3}{10}$		16.	$\dfrac{4}{5}$
3.	$\dfrac{7}{10}$		17.	$\dfrac{3}{100}$
4.	$\dfrac{9}{10}$		18.	$\dfrac{8}{100}$
5.	$\dfrac{33}{100}$		19.	$\dfrac{1}{25}$
6.	$\dfrac{69}{100}$		20.	$\dfrac{3}{25}$
7.	$\dfrac{83}{100}$		21.	$\dfrac{5}{50}$
8.	$\dfrac{91}{100}$		22.	$\dfrac{10}{25}$
9.	$\dfrac{75}{100}$		23.	$\dfrac{12}{25}$
10.	$\dfrac{3}{4}$		24.	$\dfrac{44}{50}$
11.	$\dfrac{2}{4}$		25.	$1\dfrac{1}{10}$
12.	$\dfrac{1}{2}$		26.	$1\dfrac{2}{5}$
13.	$\dfrac{1}{4}$		27.	$3\dfrac{1}{4}$
14.	$\dfrac{1}{5}$		28.	$7\dfrac{7}{10}$

524 A Express as a decimal. Second Half

1.	$\dfrac{1}{10}$		15.	$\dfrac{3}{5}$	
2.	$\dfrac{2}{10}$		16.	$\dfrac{4}{5}$	
3.	$\dfrac{5}{10}$		17.	$\dfrac{7}{100}$	
4.	$\dfrac{7}{10}$		18.	$\dfrac{8}{100}$	
5.	$\dfrac{41}{100}$		19.	$\dfrac{1}{25}$	
6.	$\dfrac{30}{100}$		20.	$\dfrac{2}{50}$	
7.	$\dfrac{96}{100}$		21.	$\dfrac{4}{25}$	
8.	$\dfrac{21}{100}$		22.	$\dfrac{10}{50}$	
9.	$\dfrac{25}{100}$		23.	$\dfrac{12}{25}$	
10.	$\dfrac{1}{4}$		24.	$\dfrac{31}{50}$	
11.	$\dfrac{2}{4}$		25.	$1\dfrac{1}{5}$	
12.	$\dfrac{1}{2}$		26.	$1\dfrac{2}{5}$	
13.	$\dfrac{3}{4}$		27.	$3\dfrac{1}{4}$	
14.	$\dfrac{1}{5}$		28.	$9\dfrac{3}{10}$	

524 B   Express as a decimal.   First Half

1.	$\dfrac{1}{10}$		15.	$\dfrac{1}{5} + \dfrac{1}{5}$
2.	$\dfrac{3}{10}$		16.	$\dfrac{3}{5} + \dfrac{1}{5}$
3.	$\dfrac{3}{10} + \dfrac{4}{10}$		17.	$\dfrac{1}{100} + \dfrac{2}{100}$
4.	$\dfrac{2}{10} + \dfrac{7}{10}$		18.	$\dfrac{5}{100} + \dfrac{3}{100}$
5.	$\dfrac{10}{100} + \dfrac{23}{100}$		19.	$\dfrac{1}{25}$
6.	$\dfrac{50}{100} + \dfrac{19}{100}$		20.	$\dfrac{2}{25} + \dfrac{1}{25}$
7.	$\dfrac{60}{100} + \dfrac{23}{100}$		21.	$\dfrac{3}{50} + \dfrac{2}{50}$
8.	$\dfrac{60}{100} + \dfrac{31}{100}$		22.	$\dfrac{8}{25} + \dfrac{2}{25}$
9.	$\dfrac{50}{100} + \dfrac{25}{100}$		23.	$\dfrac{7}{25} + \dfrac{5}{25}$
10.	$\dfrac{2}{4} + \dfrac{1}{4}$		24.	$\dfrac{39}{50} + \dfrac{5}{50}$
11.	$\dfrac{1}{4} + \dfrac{1}{4}$		25.	$\dfrac{11}{10}$
12.	$\dfrac{1}{2}$		26.	$\dfrac{7}{5}$
13.	$\dfrac{1}{8} + \dfrac{1}{8}$		27.	$\dfrac{13}{4}$
14.	$\dfrac{1}{5}$		28.	$\dfrac{77}{10}$

524 B		**Express as a decimal.**		**Second Half**

1.	$\dfrac{1}{10}$	15.	$\dfrac{2}{5} + \dfrac{1}{5}$
2.	$\dfrac{2}{10}$	16.	$\dfrac{3}{5} + \dfrac{1}{5}$
3.	$\dfrac{1}{10} + \dfrac{4}{10}$	17.	$\dfrac{4}{100} + \dfrac{3}{100}$
4.	$\dfrac{2}{10} + \dfrac{5}{10}$	18.	$\dfrac{5}{100} + \dfrac{3}{100}$
5.	$\dfrac{10}{100} + \dfrac{31}{100}$	19.	$\dfrac{1}{25}$
6.	$\dfrac{14}{100} + \dfrac{16}{100}$	20.	$\dfrac{1}{50} + \dfrac{1}{50}$
7.	$\dfrac{73}{100} + \dfrac{23}{100}$	21.	$\dfrac{3}{25} + \dfrac{1}{25}$
8.	$\dfrac{15}{100} + \dfrac{6}{100}$	22.	$\dfrac{3}{50} + \dfrac{7}{50}$
9.	$\dfrac{10}{100} + \dfrac{15}{100}$	23.	$\dfrac{7}{25} + \dfrac{5}{25}$
10.	$\dfrac{1}{4}$	24.	$\dfrac{29}{50} + \dfrac{2}{50}$
11.	$\dfrac{1}{2}$	25.	$\dfrac{6}{5}$
12.	$\dfrac{1}{4} + \dfrac{1}{4}$	26.	$\dfrac{7}{5}$
13.	$\dfrac{3}{8} + \dfrac{3}{8}$	27.	$\dfrac{13}{4}$
14.	$\dfrac{1}{5}$	28.	$\dfrac{93}{10}$

	524 A & B		Answer Sheet		First Half

1.	0.1	15.	0.4
2.	0.3	16.	0.8
3.	0.7	17.	0.03
4.	0.9	18.	0.08
5.	0.33	19.	0.04
6.	0.69	20.	0.12
7.	0.83	21.	0.1
8.	0.91	22.	0.4
9.	0.75	23.	0.48
10.	0.75	24.	0.88
11.	0.5	25.	1.1
12.	0.5	26.	1.4
13.	0.25	27.	3.25
14.	0.2	28.	7.7

524 A & B		Answer Sheet		Second Half

1.	0.1	15.	0.6
2.	0.2	16.	0.8
3.	0.5	17.	0.07
4.	0.7	18.	0.08
5.	0.41	19.	0.04
6.	0.3	20.	0.04
7.	0.96	21.	0.16
8.	0.21	22.	0.2
9.	0.25	23.	0.48
10.	0.25	24.	0.62
11.	0.5	25.	1.2
12.	0.5	26.	1.4
13.	0.75	27.	3.25
14.	0.2	28.	9.3

Math Sprints 5

Express as a fraction in simplest form.

1.	0.5	13.	0.42
2.	0.1	14.	0.6
3.	0.3	15.	0.86
4.	0.7	16.	0.05
5.	0.9	17.	0.15
6.	0.2	18.	0.35
7.	0.4	19.	1.25
8.	0.5	20.	2.5
9.	0.25	21.	3.23
10.	0.75	22.	5.05
11.	0.91	23.	0.001
12.	0.23	24.	0.002

525 A		Express as a fraction in simplest form.		Second Half

1.	0.1	13.	0.22
2.	0.7	14.	0.6
3.	0.5	15.	0.46
4.	0.9	16.	0.05
5.	0.25	17.	0.15
6.	0.2	18.	0.35
7.	0.5	19.	1.5
8.	0.4	20.	7.25
9.	0.25	21.	5.07
10.	0.75	22.	3.15
11.	0.71	23.	0.001
12.	0.37	24.	0.004

525 B	Express as a fraction in simplest form.		First Half
1.	0.5	13.	0.21 + 0.21
2.	0.1	14.	0.35 + 0.25
3.	0.1 + 0.2	15.	0.76 + 0.1
4.	0.5 + 0.2	16.	0.05
5.	0.3 + 0.6	17.	0.1 + 0.05
6.	0.1 + 0.1	18.	0.15 + 0.2
7.	0.1 + 0.3	19.	1.2 + 0.05
8.	0.25 + 0.25	20.	2.25 + 0.25
9.	0.25	21.	3.11 + 0.12
10.	0.5 + 0.25	22.	2.03 + 3.02
11.	0.89 + 0.02	23.	0.001
12.	0.19 + 0.04	24.	0.002

525 B		Express as a fraction in simplest form.		Second Half
1.	0.1		13.	0.11 + 0.11
2.	0.7		14.	0.35 + 0.25
3.	0.3 + 0.2		15.	0.36 + 0.1
4.	0.4 + 0.5		16.	0.05
5.	0.25		17.	0.09 + 0.06
6.	0.1 + 0.1		18.	0.25 + 0.1
7.	0.15 + 0.35		19.	0.9 + 0.6
8.	0.4		20.	7 + 0.25
9.	0.2 + 0.05		21.	5.03 + 0.04
10.	0.5 + 0.25		22.	2.05 + 1.1
11.	0.69 + 0.02		23.	0.001
12.	0.29 + 0.08		24.	0.004

1.	$\dfrac{1}{2}$	13.	$\dfrac{21}{50}$
2.	$\dfrac{1}{10}$	14.	$\dfrac{3}{5}$
3.	$\dfrac{3}{10}$	15.	$\dfrac{43}{50}$
4.	$\dfrac{7}{10}$	16.	$\dfrac{1}{20}$
5.	$\dfrac{9}{10}$	17.	$\dfrac{3}{20}$
6.	$\dfrac{1}{5}$	18.	$\dfrac{7}{20}$
7.	$\dfrac{2}{5}$	19.	$1\dfrac{1}{4}$
8.	$\dfrac{1}{2}$	20.	$2\dfrac{1}{2}$
9.	$\dfrac{1}{4}$	21.	$3\dfrac{23}{100}$
10.	$\dfrac{3}{4}$	22.	$5\dfrac{1}{20}$
11.	$\dfrac{91}{100}$	23.	$\dfrac{1}{1000}$
12.	$\dfrac{23}{100}$	24.	$\dfrac{1}{500}$

Math Sprints 5

1.	$\dfrac{1}{10}$	13.	$\dfrac{11}{50}$
2.	$\dfrac{7}{10}$	14.	$\dfrac{3}{5}$
3.	$\dfrac{1}{2}$	15.	$\dfrac{23}{50}$
4.	$\dfrac{9}{10}$	16.	$\dfrac{1}{20}$
5.	$\dfrac{1}{4}$	17.	$\dfrac{3}{20}$
6.	$\dfrac{1}{5}$	18.	$\dfrac{7}{20}$
7.	$\dfrac{1}{2}$	19.	$1\dfrac{1}{2}$
8.	$\dfrac{2}{5}$	20.	$7\dfrac{1}{4}$
9.	$\dfrac{1}{4}$	21.	$5\dfrac{7}{100}$
10.	$\dfrac{3}{4}$	22.	$3\dfrac{3}{20}$
11.	$\dfrac{71}{100}$	23.	$\dfrac{1}{1000}$
12.	$\dfrac{37}{100}$	24.	$\dfrac{1}{250}$

526 A Express as a percentage. First Half

1.	$\dfrac{27}{100}$	13.	$\dfrac{8}{20}$
2.	$\dfrac{38}{100}$	14.	$\dfrac{1}{2}$
3.	$\dfrac{90}{100}$	15.	$\dfrac{9}{20}$
4.	$\dfrac{63}{100}$	16.	$\dfrac{4}{10}$
5.	$\dfrac{45}{100}$	17.	$\dfrac{14}{20}$
6.	$\dfrac{75}{100}$	18.	$\dfrac{2}{25}$
7.	$\dfrac{4}{10}$	19.	$\dfrac{7}{25}$
8.	$\dfrac{7}{10}$	20.	$\dfrac{1}{4}$
9.	$\dfrac{3}{10}$	21.	$\dfrac{10}{25}$
10.	$\dfrac{5}{10}$	22.	$\dfrac{11}{25}$
11.	$\dfrac{2}{10}$	23.	$\dfrac{4}{5}$
12.	$\dfrac{3}{20}$	24.	$\dfrac{3}{5}$

Math Sprints 5

Express as a percentage.

1.	$\dfrac{29}{100}$		13.	$\dfrac{1}{20}$
2.	$\dfrac{72}{100}$		14.	$\dfrac{1}{2}$
3.	$\dfrac{1}{100}$		15.	$\dfrac{9}{20}$
4.	$\dfrac{94}{100}$		16.	$\dfrac{6}{10}$
5.	$\dfrac{23}{100}$		17.	$\dfrac{4}{20}$
6.	$\dfrac{75}{100}$		18.	$\dfrac{1}{25}$
7.	$\dfrac{8}{10}$		19.	$\dfrac{9}{25}$
8.	$\dfrac{7}{10}$		20.	$\dfrac{3}{4}$
9.	$\dfrac{9}{10}$		21.	$\dfrac{10}{25}$
10.	$\dfrac{5}{10}$		22.	$\dfrac{10}{20}$
11.	$\dfrac{2}{10}$		23.	$\dfrac{4}{5}$
12.	$\dfrac{1}{5}$		24.	$\dfrac{3}{5}$

526 B		**Express as a percentage.**		**First Half**

1.	$\dfrac{27}{100}$		13.	$\dfrac{3}{20} + \dfrac{5}{20}$
2.	$\dfrac{38}{100}$		14.	$\dfrac{4}{10} + \dfrac{2}{20}$
3.	$\dfrac{80}{100} + \dfrac{10}{100}$		15.	$\dfrac{4}{10} + \dfrac{1}{20}$
4.	$\dfrac{60}{100} + \dfrac{3}{100}$		16.	$\dfrac{2}{5}$
5.	$\dfrac{39}{100} + \dfrac{6}{100}$		17.	$\dfrac{6}{20} + \dfrac{8}{20}$
6.	$\dfrac{3}{4}$		18.	$\dfrac{2}{25}$
7.	$\dfrac{3}{10} + \dfrac{1}{10}$		19.	$\dfrac{4}{25} + \dfrac{3}{25}$
8.	$\dfrac{3}{10} + \dfrac{4}{10}$		20.	$\dfrac{2}{8}$
9.	$\dfrac{2}{10} + \dfrac{1}{10}$		21.	$\dfrac{4}{25} + \dfrac{6}{25}$
10.	$\dfrac{1}{2}$		22.	$\dfrac{5}{25} + \dfrac{6}{25}$
11.	$\dfrac{2}{10}$		23.	$\dfrac{3}{5} + \dfrac{2}{10}$
12.	$\dfrac{1}{20} + \dfrac{2}{20}$		24.	$\dfrac{2}{10} + \dfrac{2}{5}$

1.	$\dfrac{29}{100}$		13.	$\dfrac{1}{20}$
2.	$\dfrac{72}{100}$		14.	$\dfrac{4}{10} + \dfrac{2}{20}$
3.	$\dfrac{1}{100}$		15.	$\dfrac{4}{10} + \dfrac{1}{20}$
4.	$\dfrac{90}{100} + \dfrac{4}{100}$		16.	$\dfrac{3}{5}$
5.	$\dfrac{19}{100} + \dfrac{4}{100}$		17.	$\dfrac{3}{20} + \dfrac{1}{20}$
6.	$\dfrac{3}{4}$		18.	$\dfrac{1}{25}$
7.	$\dfrac{3}{10} + \dfrac{5}{10}$		19.	$\dfrac{6}{25} + \dfrac{3}{25}$
8.	$\dfrac{3}{10} + \dfrac{4}{10}$		20.	$\dfrac{6}{8}$
9.	$\dfrac{8}{10} + \dfrac{1}{10}$		21.	$\dfrac{4}{25} + \dfrac{6}{25}$
10.	$\dfrac{1}{2}$		22.	$\dfrac{15}{30}$
11.	$\dfrac{2}{10}$		23.	$\dfrac{3}{5} + \dfrac{2}{10}$
12.	$\dfrac{1}{5}$		24.	$\dfrac{2}{10} + \dfrac{2}{5}$

1.	27%	13.	40%
2.	38%	14.	50%
3.	90%	15.	45%
4.	63%	16.	40%
5.	45%	17.	70%
6.	75%	18.	8%
7.	40%	19.	28%
8.	70%	20.	25%
9.	30%	21.	40%
10.	50%	22.	44%
11.	20%	23.	80%
12.	15%	24.	60%

Math Sprints 5

1.	29%	13.	5%
2.	72%	14.	50%
3.	1%	15.	45%
4.	94%	16.	60%
5.	23%	17.	20%
6.	75%	18.	4%
7.	80%	19.	36%
8.	70%	20.	75%
9.	90%	21.	40%
10.	50%	22.	50%
11.	20%	23.	80%
12.	20%	24.	60%

527 A   Find the value.   First Half

1.	1% of 100	12.	2% of 40
2.	10% of 100	13.	22% of 40
3.	10% of 10	14.	10% of 100
4.	10% of 20	15.	10% of 200
5.	5% of 20	16.	50% of 200
6.	15% of 20	17.	15% of 200
7.	10% of 40	18.	1% of 200
8.	20% of 40	19.	10% of 50
9.	10% of 10	20.	15% of 60
10.	1% of 10	21.	12% of 50
11.	1% of 40	22.	18% of 50

527 A   Find the value.   Second Half

1.	2% of 100		12.	2% of 90
2.	15% of 100		13.	30% of 40
3.	10% of 10		14.	10% of 100
4.	20% of 10		15.	10% of 200
5.	10% of 20		16.	50% of 400
6.	15% of 20		17.	15% of 400
7.	10% of 60		18.	1% of 400
8.	20% of 40		19.	10% of 50
9.	10% of 100		20.	15% of 60
10.	1% of 10		21.	12% of 50
11.	1% of 90		22.	18% of 50

527 B		Find the value.		First Half

1.	1% of 100	12.	2% of 40	
2.	50% of 20	13.	11% of 80	
3.	10% of 10	14.	10% of 100	
4.	10% of 20	15.	10% of 200	
5.	5% of 20	16.	25% of 400	
6.	15% of 20	17.	15% of 200	
7.	20% of 20	18.	5% of 40	
8.	40% of 20	19.	20% of 25	
9.	10% of 10	20.	20% of 45	
10.	1% of 10	21.	24% of 25	
11.	1% of 40	22.	36% of 25	

Find the value.

1.	2% of 100		12.	2% of 90
2.	50% of 30		13.	15% of 80
3.	10% of 10		14.	10% of 100
4.	20% of 10		15.	10% of 200
5.	50% of 4		16.	25% of 800
6.	15% of 20		17.	15% of 400
7.	20% of 30		18.	5% of 80
8.	40% of 20		19.	20% of 25
9.	20% of 50		20.	20% of 45
10.	1% of 10		21.	24% of 25
11.	1% of 90		22.	36% of 25

527 A & B		Answer Sheet		First Half

1.	1	12.	0.8
2.	10	13.	8.8
3.	1	14.	10
4.	2	15.	20
5.	1	16.	100
6.	3	17.	30
7.	4	18.	2
8.	8	19.	5
9.	1	20.	9
10.	0.1	21.	6
11.	0.4	22.	9

527 A & B   Answer Sheet   Second Half

1.	2	12.	1.8
2.	15	13.	12
3.	1	14.	10
4.	2	15.	20
5.	2	16.	200
6.	3	17.	60
7.	6	18.	4
8.	8	19.	5
9.	10	20.	9
10.	0.1	21.	6
11.	0.9	22.	9

528 A   Find the average.   First Half

1.	1, 2, and 3	12.	5, 10, and 15
2.	5, 6, and 7	13.	25, 30, and 20
3.	12, 13, and 14	14.	3, 6, and 9
4.	7, 8, and 9	15.	12, 15, and 18
5.	4, 5, and 6	16.	1, 4, and 7
6.	9, 8, and 10	17.	4, 4, and 1
7.	3, 5, and 4	18.	2, 5, and 8
8.	12, 11, and 13	19.	13, 7, and 1
9.	1, 3, and 5	20.	2, 14, and 14
10.	2, 4, and 6	21.	3, 12, and 18
11.	14, 16, and 18	22.	11, 6, and 19

528 A   Find the average.   Second Half

1.	2, 3, and 4	12.	5, 10, and 15
2.	4, 5, and 6	13.	25, 30, and 20
3.	12, 13, and 14	14.	8, 5, and 11
4.	10, 11, and 12	15.	14, 4, and 9
5.	9, 10, and 11	16.	1, 4, and 7
6.	9, 8, and 10	17.	4, 4, and 1
7.	2, 4, and 3	18.	6, 6, and 3
8.	12, 11, and 13	19.	1, 6, and 11
9.	2, 4, and 6	20.	2, 14, and 14
10.	2, 6, and 4	21.	3, 12, and 18
11.	7, 9, and 11	22.	10, 12, and 20

528 B   Find the average.   First Half

1.	3, 2, and 1	12.	15, 5, and 10
2.	7, 5, and 6	13.	25, 30, and 20
3.	14, 12, and 13	14.	9, 3, and 6
4.	9, 7, and 8	15.	18, 15, and 12
5.	6, 4, and 5	16.	1, 5, and 6
6.	7, 9, and 11	17.	4, 1, and 4
7.	2, 4, and 6	18.	8, 2, and 5
8.	10, 12, and 14	19.	18, 1, and 2
9.	5, 1, and 3	20.	15, 2, and 13
10.	6, 4, and 2	21.	18, 11, and 4
11.	16, 18, and 14	22.	11, 6, and 19

528 B   Find the average.   Second Half

1.	4, 3, and 2	12.	15, 5, and 10
2.	4, 5, and 6	13.	25, 30, and 20
3.	14, 12, and 13	14.	22, 1, and 1
4.	12, 10, and 11	15.	11, 8, and 8
5.	9, 11, and 10	16.	1, 5, and 6
6.	7, 9, and 11	17.	4, 1, and 4
7.	1, 3, and 5	18.	8, 2, and 5
8.	10, 12, and 14	19.	15, 1, and 2
9.	2, 6, and 4	20.	15, 2, and 13
10.	6, 4, and 2	21.	18, 11, and 4
11.	9, 11, and 7	22.	10, 20, and 12

528 A & B	Answer Sheet	First Half

1.	2	12.	10
2.	6	13.	25
3.	13	14.	6
4.	8	15.	15
5.	5	16.	4
6.	9	17.	3
7.	4	18.	5
8.	12	19.	7
9.	3	20.	10
10.	4	21.	11
11.	16	22.	12

Math Sprints 5

1.	3		12.	10
2.	5		13.	25
3.	13		14.	8
4.	11		15.	9
5.	10		16.	4
6.	9		17.	3
7.	3		18.	5
8.	12		19.	6
9.	4		20.	10
10.	4		21.	11
11.	9		22.	14

Math Sprints 5

529 A	Find the measure of the 3rd interior angle of a triangle.		First Half

1.	∠a = 60°, ∠b = 60°	13.	∠a = 62°, ∠b = 58°
2.	∠a = 70°, ∠b = 60°	14.	∠a = 102°, ∠b = 8°
3.	∠a = 80°, ∠b = 60°	15.	∠a = 4°, ∠b = 7°
4.	∠a = 80°, ∠b = 70°	16.	∠a = 22°, ∠b = 33°
5.	∠a = 50°, ∠b = 60°	17.	∠a = 61°, ∠b = 60°
6.	∠a = 25°, ∠b = 25°	18.	∠a = 61°, ∠b = 62°
7.	∠a = 95°, ∠b = 10°	19.	∠a = 98°, ∠b = 20°
8.	∠a = 95°, ∠b = 35°	20.	∠a = 97°, ∠b = 19°
9.	∠a = 99°, ∠b = 20°	21.	∠a = 101°, ∠b = 20°
10.	∠a = 32°, ∠b = 48°	22.	∠a = 102°, ∠b = 20°
11.	∠a = 46°, ∠b = 74°	23.	∠a = 96°, ∠b = 42°
12.	∠a = 56°, ∠b = 74°	24.	∠a = 65°, ∠b = 43°

529 A Find the measure of the 3rd interior angle of a triangle. Second Half

1.	∠a = 60°, ∠b = 60°	13.	∠a = 61°, ∠b = 59°
2.	∠a = 50°, ∠b = 60°	14.	∠a = 102°, ∠b = 8°
3.	∠a = 40°, ∠b = 60°	15.	∠a = 4°, ∠b = 7°
4.	∠a = 30°, ∠b = 70°	16.	∠a = 22°, ∠b = 33°
5.	∠a = 70°, ∠b = 60°	17.	∠a = 61°, ∠b = 60°
6.	∠a = 25°, ∠b = 25°	18.	∠a = 61°, ∠b = 62°
7.	∠a = 90°, ∠b = 70°	19.	∠a = 98°, ∠b = 20°
8.	∠a = 95°, ∠b = 35°	20.	∠a = 97°, ∠b = 19°
9.	∠a = 99°, ∠b = 60°	21.	∠a = 101°, ∠b = 59°
10.	∠a = 31°, ∠b = 49°	22.	∠a = 100°, ∠b = 61°
11.	∠a = 45°, ∠b = 75°	23.	∠a = 95°, ∠b = 43°
12.	∠a = 46°, ∠b = 84°	24.	∠a = 65°, ∠b = 43°

529 B   Find the measure of the 3rd interior angle of a triangle.   First Half

1.	∠a = 60°, ∠b = 60°	13.	∠a = 26°, ∠b = 94°
2.	∠a = 70°, ∠b = 60°	14.	∠a = 92°, ∠b = 18°
3.	∠a = 75°, ∠b = 65°	15.	∠a = 8°, ∠b = 3°
4.	∠a = 80°, ∠b = 70°	16.	∠a = 23°, ∠b = 32°
5.	∠a = 45°, ∠b = 65°	17.	∠a = 62°, ∠b = 59°
6.	∠a = 15°, ∠b = 35°	18.	∠a = 64°, ∠b = 59°
7.	∠a = 85°, ∠b = 20°	19.	∠a = 99°, ∠b = 19°
8.	∠a = 85°, ∠b = 45°	20.	∠a = 98°, ∠b = 18°
9.	∠a = 99°, ∠b = 20°	21.	∠a = 102°, ∠b = 19°
10.	∠a = 33°, ∠b = 47°	22.	∠a = 103°, ∠b = 19°
11.	∠a = 36°, ∠b = 84°	23.	∠a = 89°, ∠b = 49°
12.	∠a = 46°, ∠b = 84°	24.	∠a = 23°, ∠b = 85°

529 B   Find the measure of the 3rd interior angle of a triangle.   Second Half

1.	$\angle a = 60°, \angle b = 60°$	13.	$\angle a = 26°, \angle b = 94°$
2.	$\angle a = 50°, \angle b = 60°$	14.	$\angle a = 82°, \angle b = 28°$
3.	$\angle a = 35°, \angle b = 65°$	15.	$\angle a = 8°, \angle b = 3°$
4.	$\angle a = 30°, \angle b = 70°$	16.	$\angle a = 21°, \angle b = 34°$
5.	$\angle a = 65°, \angle b = 65°$	17.	$\angle a = 63°, \angle b = 58°$
6.	$\angle a = 15°, \angle b = 35°$	18.	$\angle a = 64°, \angle b = 59°$
7.	$\angle a = 85°, \angle b = 75°$	19.	$\angle a = 99°, \angle b = 19°$
8.	$\angle a = 85°, \angle b = 45°$	20.	$\angle a = 98°, \angle b = 18°$
9.	$\angle a = 99°, \angle b = 60°$	21.	$\angle a = 103°, \angle b = 57°$
10.	$\angle a = 32°, \angle b = 48°$	22.	$\angle a = 103°, \angle b = 58°$
11.	$\angle a = 46°, \angle b = 74°$	23.	$\angle a = 79°, \angle b = 59°$
12.	$\angle a = 39°, \angle b = 91°$	24.	$\angle a = 33°, \angle b = 75°$

1.	60°	13.	60°
2.	50°	14.	70°
3.	40°	15.	169°
4.	30°	16.	125°
5.	70°	17.	59°
6.	130°	18.	57°
7.	75°	19.	62°
8.	50°	20.	64°
9.	61°	21.	59°
10.	100°	22.	58°
11.	60°	23.	42°
12.	50°	24.	72°

	529 A & B		Answer Sheet		Second Half

1.	60°	13.	60°
2.	70°	14.	70°
3.	80°	15.	169°
4.	80°	16.	125°
5.	50°	17.	59°
6.	130°	18.	57°
7.	20°	19.	62°
8.	50°	20.	64°
9.	21°	21.	20°
10.	100°	22.	19°
11.	60°	23.	42°
12.	50°	24.	72°

530 A   Find the measure of the 3rd interior angle of a triangle.   **First Half**

1.	∠a = 60°, ∠b = 60°	13.	∠a = 39°, ∠b = 82°
2.	∠a = 80°, ∠b = 60°	14.	∠a = 39°, ∠b = 84°
3.	∠a = 90°, ∠b = 60°	15.	∠a = 103°, ∠b = 32°
4.	∠a = 60°, ∠b = 100°	16.	∠a = 105°, ∠b = 32°
5.	∠a = 65°, ∠b = 100°	17.	∠a = 105°, ∠b = 31°
6.	∠a = 10°, ∠b = 10°	18.	∠a = 76°, ∠b = 74°
7.	∠a = 20°, ∠b = 10°	19.	∠a = 87°, ∠b = 74°
8.	∠a = 19°, ∠b = 10°	20.	∠a = 33°, ∠b = 68°
9.	∠a = 11°, ∠b = 19°	21.	∠a = 35°, ∠b = 68°
10.	∠a = 54°, ∠b = 46°	22.	∠a = 68°, ∠b = 55°
11.	∠a = 23°, ∠b = 67°	23.	∠a = 92°, ∠b = 87°
12.	∠a = 38°, ∠b = 82°	24.	∠a = 72°, ∠b = 85°

530 A   Find the measure of the 3rd interior angle of a triangle.   Second Half

1.	∠a = 90°, ∠b = 10°	13.	∠a = 39°, ∠b = 82°
2.	∠a = 80°, ∠b = 20°	14.	∠a = 38°, ∠b = 58°
3.	∠a = 90°, ∠b = 60°	15.	∠a = 113°, ∠b = 22°
4.	∠a = 40°, ∠b = 100°	16.	∠a = 115°, ∠b = 22°
5.	∠a = 65°, ∠b = 100°	17.	∠a = 105°, ∠b = 31°
6.	∠a = 10°, ∠b = 20°	18.	∠a = 66°, ∠b = 40°
7.	∠a = 40°, ∠b = 10°	19.	∠a = 87°, ∠b = 74°
8.	∠a = 39°, ∠b = 10°	20.	∠a = 23°, ∠b = 78°
9.	∠a = 11°, ∠b = 19°	21.	∠a = 35°, ∠b = 68°
10.	∠a = 51°, ∠b = 49°	22.	∠a = 68°, ∠b = 55°
11.	∠a = 13°, ∠b = 77°	23.	∠a = 93°, ∠b = 86°
12.	∠a = 18°, ∠b = 102°	24.	∠a = 72°, ∠b = 85°

530 B   Find the measure of the 3rd interior angle of a triangle.   First Half

1.	∠a = 60°, ∠b = 60°	13.	∠a = 49°, ∠b = 72°
2.	∠a = 75°, ∠b = 65°	14.	∠a = 39°, ∠b = 84°
3.	∠a = 85°, ∠b = 65°	15.	∠a = 93°, ∠b = 42°
4.	∠a = 85°, ∠b = 75°	16.	∠a = 94°, ∠b = 43°
5.	∠a = 75°, ∠b = 90°	17.	∠a = 84°, ∠b = 52°
6.	∠a = 5°, ∠b = 15°	18.	∠a = 36°, ∠b = 114°
7.	∠a = 15°, ∠b = 15°	19.	∠a = 47°, ∠b = 114°
8.	∠a = 14°, ∠b = 15°	20.	∠a = 24°, ∠b = 77°
9.	∠a = 16°, ∠b = 14°	21.	∠a = 35°, ∠b = 68°
10.	∠a = 64°, ∠b = 36°	22.	∠a = 59°, ∠b = 64°
11.	∠a = 33°, ∠b = 57°	23.	∠a = 92°, ∠b = 87°
12.	∠a = 47°, ∠b = 73°	24.	∠a = 61°, ∠b = 96°

530 B   Find the measure of the 3rd interior angle of a triangle.   Second Half

1.	∠a = 90°, ∠b = 10°	13.	∠a = 49°, ∠b = 72°
2.	∠a = 40°, ∠b = 60°	14.	∠a = 39°, ∠b = 57°
3.	∠a = 75°, ∠b = 75°	15.	∠a = 83°, ∠b = 52°
4.	∠a = 65°, ∠b = 75°	16.	∠a = 84°, ∠b = 53°
5.	∠a = 75°, ∠b = 90°	17.	∠a = 84°, ∠b = 52°
6.	∠a = 5°, ∠b = 25°	18.	∠a = 33°, ∠b = 73°
7.	∠a = 25°, ∠b = 25°	19.	∠a = 47°, ∠b = 114°
8.	∠a = 24°, ∠b = 25°	20.	∠a = 34°, ∠b = 67°
9.	∠a = 16°, ∠b = 14°	21.	∠a = 35°, ∠b = 68°
10.	∠a = 74°, ∠b = 26°	22.	∠a = 59°, ∠b = 64°
11.	∠a = 34°, ∠b = 56°	23.	∠a = 93°, ∠b = 86°
12.	∠a = 43°, ∠b = 77°	24.	∠a = 61°, ∠b = 96°

                              © SingaporeMath.com Inc.

530 A & B		Answer Sheet		First Half

1.	60°	13.	59°
2.	40°	14.	57°
3.	30°	15.	45°
4.	20°	16.	43°
5.	15°	17.	44°
6.	160°	18.	30°
7.	150°	19.	19°
8.	151°	20.	79°
9.	150°	21.	77°
10.	80°	22.	57°
11.	90°	23.	1°
12.	60°	24.	23°

Math Sprints 5

1.	80°	13.	59°
2.	80°	14.	84°
3.	30°	15.	45°
4.	40°	16.	43°
5.	15°	17.	44°
6.	150°	18.	74°
7.	130°	19.	19°
8.	131°	20.	79°
9.	150°	21.	77°
10.	80°	22.	57°
11.	90°	23.	1°
12.	60°	24.	23°